Computational Rheology for Pipeline and Annular Flow

Computational Rheology for Pipeline and Annular Flow

Editor

Sudhir Chitnis

scitus
academics

Computational Rheology for Pipeline and Annular Flow

Edited by **Sudhir Chitnis**

Printed in 2017

ISBN: 978-1-68117-347-4

Library of Congress Control Number: 2015939260

© 2016 by

SCITUS Academics LLC,
616, Corporate Way, Suite 2, 4766,
Valley Cottage, NY 10989

www.scitusacademics.com

Notice

Reasonable efforts have been made to publish reliable data and views articulated in the chapters are those of the individual contributors, and not necessarily those of the editors or publishers. Editors or publishers are not responsible for the accuracy of the information in the published chapters or consequences of their use. The publisher believes no responsibility for any damage or grievance to the persons or property arising out of the use of any materials, instructions, methods or thoughts in the book. The editors and the publisher have attempted to trace the copyright holders of all material reproduced in this publication and apologize to copyright holders if permission has not been obtained. If any copyright holder has not been acknowledged, please write to us so we may rectify.

Contents

Preface

Computational Rheology for Pipeline and Annular Flow develops and applies modern analytical and computational finite difference methods for solving flow problems in drilling and production. It also provides valuable insights into flow assurance analysis in subsea pipeline design. Using modeling techniques that simulate the motion of non-Newtonian fluids this book presents proven annular flow methodologies for cuttings transport and stuck pipe analysis based on detailed experimental data obtained from highly deviated and horizontal wells. These methods are applied for highly eccentric borehole geometries to the design of pipeline bundles in subsea applications, where such annular configurations arise in velocity and thermal modeling applications.

Editor

Numerical Analysis of Heavy Oil-Water Flow and Leak Detection in Vertical Pipeline

João Victor Nunes de Sousa[1], Cristiane Holanda Sodré[2], Antonio Gilson Barbosa de Lima[1], and Severino Rodrigues de Farias Neto[3]

[1]Department of Mechanical Engineering, Center of Sciences and Technology, Federal University of Campina Grande, Campina Grande, Brazil

[2]Department of Chemical Engineering, Technology Center, Federal University of Alagoas, Maceió, Brazil

[3]Department of Chemical Engineering, Center of Sciences and Technology, Federal University of Campina Grande, Campina Grande, Brazil

ABSTRACT

Pipeline is a conventional, efficient and economic way for oil transportations. The use of a good system for detecting and locating leaks in pipeline contribute significantly to operational safety and cost saving in petroleum industry. This paper aims to study the heavy oil-water flow in vertical ducts including leakage. A transient numerical analysis, using the ANSYS-CFX®11.0 commercial software is performed. The mathematical modeling considers the effect of drag and gravitational forces between the phases and turbulent flow. Mass flow rate of the phases in the leaking orifice, the pressure drop as a function of the time and the velocity distributions are presented and discussed. We can conclude that volumetric fraction of phases and fluid mixture velocity affect pressure drop and mass flow rate at the leak hole.

INTRODUCTION

The activity of oil production is subject to high risks. Even the petroleum industry running preventive measures, there is always the possibility of failure, making the industrial plants susceptible to operational accidents with loss of fluid to environment, causing great ecological, social and economic damages, with delay in oil production. A proper supervisory system must be capable of detecting leaks in oil installations, enabling immediate action to reduce the impacts of accidents and contributing significantly to operational safety. The simultaneous flow of two immiscible liquids in vertical pipes is encountered in different industrials processes and particularly in the petroleum industry [1].

Because of importance, many authors have focused their researches in methods of leak detection in pipes on oil production and transport [2-4].

However, in different applications, including oil transportation, accurate locations of the leaks is still very difficult. In present day various leak detection techniques based in the negative pressure wave, acoustic sensors, satellite surveillance, mass and volume balance, analytical model-based method, among others, has been applied. All these methods are based in process variables such as pressure, mass and volumetric flow rates and temperature [5].

According to Dong et al. [6], the negative pressure method, which supplies high leak sensitivity and availability, is a relatively better method among them. Unfortunately this method has a high possibility of false alarm if there are some strong raises in the pressure measurement records or if the leak is small (0.5% of nominal flow) [4,5].

Thus, this paper aims to numerically study the hydrodynamic of heavy oil-water flow in a vertical pipe having a small leak, which is much more difficult to detect by conventional systems [7]. The interest in heavy oil is in fact that recent studies indicate that in 2025 this kind of oil will be the main source of fossil energy in the world [8].

METHODOLOGY

The Geometry and Grid

The study domain (Figure 1) consists of a vertical pipe with 800 cm (8 m) of length, with a constant circular section 15 cm diameter. To simulate the leakage, the pipe has a circular hole, with 0.6 cm diameter, located at the midpoint of the length of pipe.

Figure 2 illustrates the mesh representing the study domain, which was built with the support of ICEMCFD® 11.0 software. This structured mesh was obtained after various refinements, and it has 327,327 hexahedral elements.

Mathematical Modeling

To investigate the multiphase flow of heavy oil-water in vertical pipes leaking, it was considered three-dimensional, transient and isothermal flow, non-homogeneous model for the fluid mixture (particle model) and homogeneous model for turbulence (k-ε model). The homogeneous model considers a single field for both phases, while in the non-homogeneous model is considered a specific field for each phase [9].

On the mathematical modeling, the index α represent the continuous phase (oil) and the index β represent the dispersed phase (water). The dispersed water phase is modeled as spherical particles.

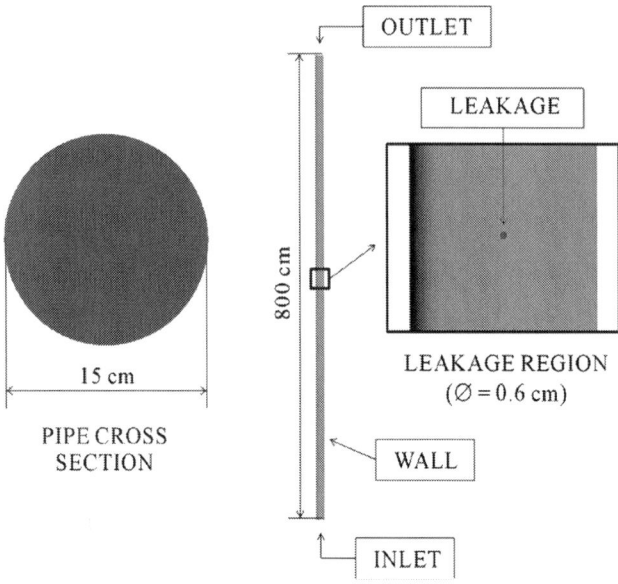

Figure 1: Considered pipe layout.

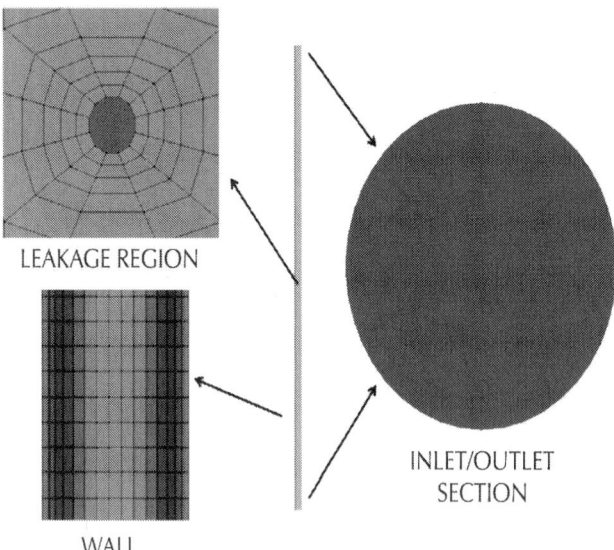

Figure 2: Studied pipe, showing different regions of the mesh.

The general equations used in this work are:

-      Continuity Equations,

$$\frac{\partial\left(f_\alpha\rho_\alpha\right)}{\partial t}+\nabla\cdot\left(f_\alpha\rho_\alpha U_\alpha\right)=0,$$

(1)

$$\frac{\partial\left(f_\beta\rho_\beta\right)}{\partial t}+\nabla\cdot\left(f_\beta\rho_\beta U_\beta\right)=0,$$

(2)

where f is the volume fraction, ρ is the density and U=(u,v,w) is the velocity vector, each corresponding to a given phase;

-    Momentum Equations,

$$\frac{\partial\left(f_\alpha\rho_\alpha U_\alpha\right)}{\partial t}+\nabla\cdot\left[f_\alpha\left(\rho_\alpha U_\alpha\otimes U_\alpha\right)\right]$$

$$=-f_\alpha\nabla p'+\nabla\cdot\left\{f_\alpha\mu_{eff,\alpha}\left[\nabla U_\alpha+\left(\nabla U_\alpha\right)^T\right]\right\}+S_{M\alpha}+D,$$

(3)

$$\frac{\partial\left(f_\beta\rho_\beta U_\beta\right)}{\partial t}+\nabla\cdot\left[f_\beta\left(\rho_\beta U_\beta\otimes U_\beta\right)\right]$$

$$=-f_\beta\nabla p'+\nabla\cdot\left\{f_\beta\mu_{eff,\beta}\left[\nabla U_\beta+\left(\nabla U_\beta\right)^T\right]\right\}+S_{M\beta}-D,$$

(4)

where p' is the modified pressure, μ_{eff} is the effective viscosity, S_M is the momentum sources due to external body forces (when gravitational forces are includes) and D is the drag force between the phases, that is modeled by the equation

$$D=\frac{1}{8}C_D\rho_\alpha A\left|U_\beta-U_\alpha\right|\left(U_\beta-U_\alpha\right),$$

(5)

where C_D is the drag coefficient and A is the interfacial area density. For Re < 1000, the drag coefficient is modeled by Schiller-Naumann model,

$$C_D=\frac{24}{Re}\left(1+0.15Re^{0.687}\right),$$

(6)

and for Re ≥ 1000, the drag coefficient is considered 0.44, where Re represents the particle Reynolds number, modeled by

$$Re = \frac{\rho_\alpha \left| U_\beta - U_\alpha \right| d_\beta}{\mu_\alpha},$$

(7)

where d is the diameter of spherical particles (3 mm). The interfacial area density, A, is modeled by the equation

$$A = \frac{6 f_\beta}{d_\beta};$$

(8)

- Kinetic Energy Equation,

$$\frac{\partial(\rho_m k)}{\partial t} + \nabla \cdot (\rho_m U_m k) = \nabla \cdot \left[\left(\mu_m + \frac{\mu_t}{C_k} \right) \nabla k \right] + P_k - \rho_m \varepsilon,$$

(9)

where ρ_m is the density, k is the kinetic energy, U_m is the velocity vector, μ_m is the viscosity, μ_t is the turbulent viscosity, P_k is the turbulence production and ε is the turbulence eddy dissipation, each corresponding to a mixture;

- Turbulence Eddy Dissipation Equation,

$$\frac{\partial(\rho_m \varepsilon)}{\partial t} + \nabla \cdot (\rho_m U_m \varepsilon)$$

$$= \nabla \cdot \left[\left(\mu_m + \frac{\mu_t}{C_{\varepsilon 1}} \right) \nabla \varepsilon \right] + \frac{\varepsilon}{k} (C_{\varepsilon 2} P_k - C_{\varepsilon 3} \rho_m \varepsilon),$$

(10)

where C_k = 1.0, $C\varepsilon_1$ = 1.3, $C\varepsilon 2$ = 1.44 and $C_{\varepsilon 3}$ = 1.92. In the k-ε model the parameters as follows

$$p' = p_{din} + \frac{2}{3} \rho_m k,$$

(11)

$$\mu_{eff,\alpha} = \mu_\alpha + \mu_t,$$

(12)

$$\mu_{eff,\beta} = \mu_\beta + \mu_t,$$
(13)

$$\rho_m = f_\alpha \rho_\alpha + f_\beta \rho_\beta,$$
(14)

$$U_m = \frac{1}{\rho}\left(f_\alpha \rho_\alpha U_\alpha + f_\beta \rho_\beta U_\beta\right),$$
(15)

$$\mu_m = f_\alpha \mu_\alpha + f_\beta \mu_\beta,$$
(16)

$$\mu_t = C_\mu \rho_m \frac{k^2}{\varepsilon},$$
(17)

$$P_k = \mu_t \nabla U \cdot \left(\nabla U + \nabla U^T\right) - \frac{2}{3}\nabla \cdot U\left(3\mu_t \nabla \cdot U + \rho k\right),$$
(18)

where p_{din} is the dynamic pressure, μ/μ is the viscosity of phase and $C_\mu = 0.09$. The dynamic pressure is calculated by the equation

$$p_{din} = \frac{1}{2}\left(\rho_\alpha f_\alpha |U_\alpha|^2 + \rho_\beta f_\beta |U_\beta|^2\right).$$
(19)

In multiphase flow, the total pressure acting in the phases, p_{tot}, is modeled by

$$p_{tot} = p_{st} + p_{din},$$
(20)

where p_{st} is a term correspondent to the static pressure.

Initial and Boundary Conditions and Fluids Properties

Initially (at time t = 0), the leak in the pipe does not exist, and the multiphase flow occurs in steady state condition. When t > 0 the leak appears abruptly in the pipeline. Tables 1 and 2 shows the initial and boundary conditions used in the simulations.

Simulations were performed with the following situations:

- Oil volume fraction at the inlet, f_o, ranging from 0.75 to 1.00 (step 0.05) and the water volume fraction at the inlet, fw, corresponds to the remaining fraction, with the velocity mixture at the inlet, U, fixed in 1.00 m/s;

Table 1: Boundary conditions for t = 0

Boundary	t = 0			
	p [Pa]	U [m/s]	f_o [-]	f_w [-]
Inlet		0.75 - 2.00	0.75 - 1.00	0.00 - 0.25
Outlet	101,325			
Leakage[a]		0		
Wall		0		

[a]Impermeable.

Table 2: Boundary conditions for t > 0

Boundary	**t>0**			
	p [Pa]	**U [m/s]**	f_o **[-]**	f_w **[-]**
Inlet		0.75 - 2.00	0.75 - 1.00	0.00 - 0.25
Outlet	101,325			
Leakage[a]	101,325			
Wall		0		

[a]Opening.

- Mixture velocity, U, ranging from 0.75 to 2.00 m/s (step 0.25), with the water volume fraction, f_w, and oil volume fraction, f_o, fixed in 0.15 and 0.85, respectively.

The adopted properties of the fluids are showed in Table 3.

Table 4 shows the considerations adopted for the numerical solver.

RESULTS AND DISCUSSION

The simulations were performed on a Intel® Core 2 Quad 2.4 GHz, 4 GB RAM and 1 Tb physical memory (HD) computer. The solving time of the studied cases ranged from 10 to 11 hours.

Figure 3 shows the total mass flow rate relationships on the leakage, m_{leak}/m_{in}, as a function of oil volume fraction at the inlet section, $f_{o,in}$, to post-leak system. We can see that the smaller the oil volume fraction, consequently, greater the water volume fraction in the mixture, thus we have a large amount of fluids exiting in the leak. This behavior is explained by the reduction of the viscosity of the fluid mixture, proportional to the water holdup contained in it. Figure 4 shows the total and oil mass flow rates in the leak, m_{leak}, as a function of the oil volume fraction at the inlet section of pipe, to post-leak system. Note that the amount of water and oil mass flow rate in leak is proportional to the holdup of phases present in the mixture. Equations (21)-(23) were obtained by fitting the bared on the numerical data.

Table 3: Fluids properties

Fluid	ρ [kg/m³]	μ [mPa·s]	τ [N/m][a]
Water	997.0	0.8899	0.02905
Oil	925.5	500	

[a]τ is the oil-water surface tension.

Table 4: Numerical simulation characteristics

Characteristic	Consideration
Flow regime	Transient
Total simulation time	0.100 s
Time step	0.001 s
Convergence criterion for mass and momentum	10^{-7} (RMS)
Advection scheme	High resolution
Transient scheme	Second order
Interpolation scheme for pressure	Trilinear
Interpolation scheme for velocity	Trilinear

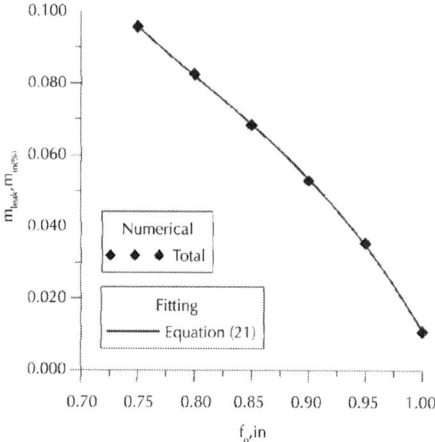

Figure 3: Total mass flow rate relationship in the leakage as a function of the oil volume fraction at the pipe inlet (U = 1.00 m/s).

$$_{ak}/m_{in} = 1.602813447 - 5.122188379\,f_{o,in}$$
$$+\,6.011162647\,f_{o,in}^2 - 2.480859856\,f_o \qquad (21)$$

$$m_{leak,total} = 262.9121767 - 834.7170159\,f_{o,in}$$
$$+\,974.6134762\,f_{o,in}^2 - 401.0288889\,f_{o,in}^3 \cdot \qquad (22)$$

$$m_{leak,oil} = 236.8741911 - 828.5900693\,f_{o,in}$$
$$+\,1036.792032\,f_{o,in}^2 - 443.2948148\,f_{o,in}^3 \cdot \qquad (23)$$

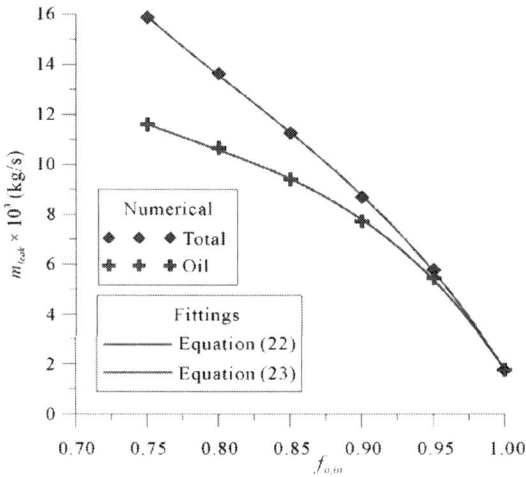

Figure 4: Total and oil mass flow rates in the leakage as a function of the oil volume fraction at the pipe inlet (U = 1.00 m/s).

Figure 5 shows the total mass flow rate relationships in leakage, m_{leak}/m_{in}, as a function of the fluid mixture velocity at the inlet, $U_{m,in}$, to post-leak system. It is noted that with increasing velocity, there is a reduction in the total mass flow rate through in the leaking. However, there is a reversal point ($U_{m,in}$ = 1.50 m/s), where by increasing the velocity of flow, we have an increased total mass flow rate in the leaking. This reversal point would be the point where the inertial forces of the flow becomes less than the forces caused by pressure differential between the inside and outside of the pipeline leaking (pressure gradients). Figure 6 shows the mass flow rate in the leak, m_{leak}, as a function of the mixture velocity at the inlet, to post-leak. It is visible that increasing the mixture velocity at the inlet, the oil mass flow in the leaking increase, and the water mass flow is practically constant. Equations (24)-(26) were obtained by fitting the bared on the numerical data.

$$m_{leak}/m_{in} = 0.2331399521 - 0.3712444597U_{m,in}$$
$$+ 0.2980465586U_{m,in}^2 - 0.1059246175U_{m,in}^3$$
$$+ 0.01427994068U_{m,in}^4 . \qquad (24)$$

$$m_{leak,total} = 12.19420476 - 8.811055026U_{m,in}$$
$$+ 9.052387302U_{m,in}^2 - 1.2139925926U_{m,in}^3. \qquad (25)$$

$$m_{leak,oil} = 10.14426714 - 7.349817778U_{m,in}$$
$$+ 7.577278095U_{m,in}^2 - 1.010542222U_{m,in}^3. \qquad (26)$$

Table 5 shows the determination coefficients, R^2, obtained in the fittings correspondents to Equations (21) to (26). These equations were obtained by the method of the least squares.

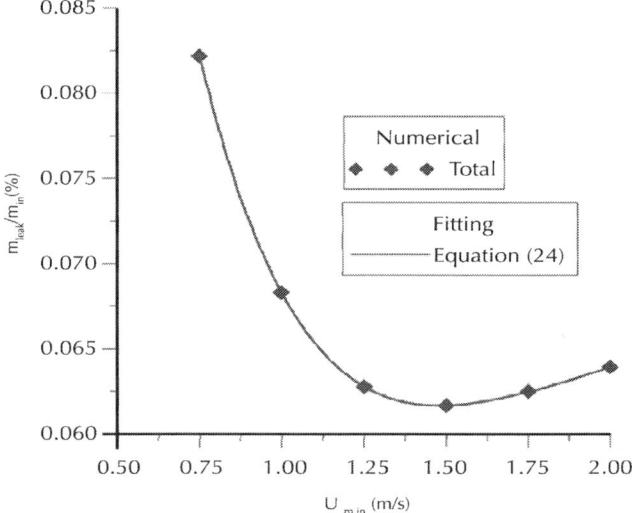

Figure 5: Total mass flow rate relationship in the leakage as a function of the fluid mixture velocity at the pipe inlet ($f_{o,in} = 0.85$).

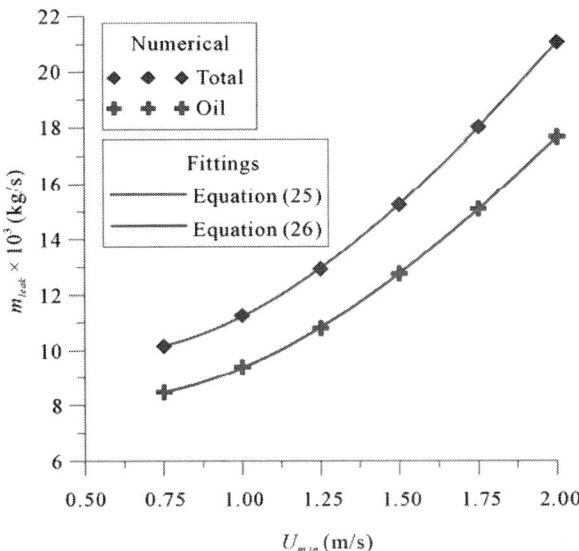

Figure 6: Total and oil mass flow rates in the leakage as a function of the fluid mixture velocity at the pipe inlet ($f_{o,in}$ = 0.85).

Figure 7 shows the oil velocity vectors at the leakage section, for some oil holdups in the mixture at the pipe inlet ($f_{o,in}$ = 0.80, 0.90 and 1.00). Note that by reducing the oil holdup, and thus, increasing the water holdup, there is a greater spread in the leakage, with a larger angle, taking as reference to the longitudinal direction of the pipe.

Figure 8 shows the velocity vectors at the leakage section, for some fluid mixture velocities at the pipe inlet. (Um,in = 1.00, 1.50 and 2.00 m/s). It is visible the small influence of the velocity in the leak spread and angle.

Table 5. Determination coefficients in the fittings

Equation	R^2
21	0.999837
22	0.999845
23	0.999652
24	0.999995

25	0.999974
26	0.999974

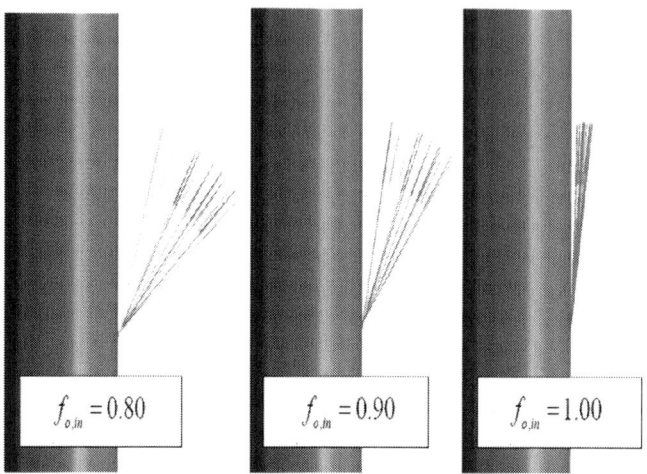

Figure 7: Velocity vectors at the leakage section, for some oil volume fractions in the mixture at the pipe inlet ($U_{m,in}$ = 1.00 m/s).

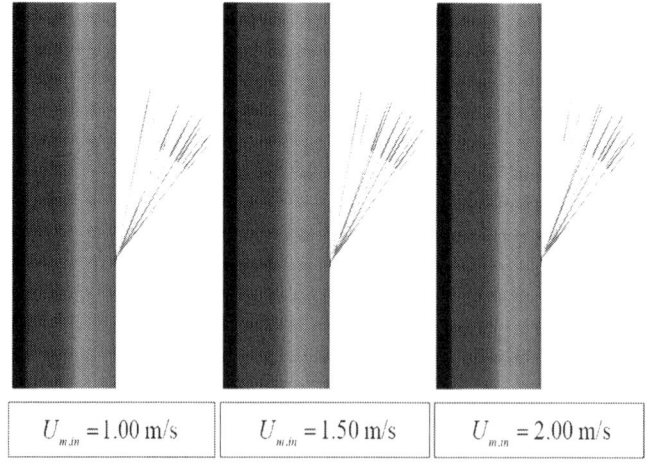

Figure 8: Velocity vectors at the leakage section, for some fluid mixture velocities at the pipe inlet ($f_{o,in}$ = 0.85).

Figure 9 shows the total pressure drop, Δp_{tot}, as a function of the time, t, for different oil volume fractions at the inlet, for a pipe section of 4 m length, where the leak is in the midpoint of this section. In all cases, it is visible reduction in the pressure drop at the initial instants of the leakage. The transient period is short (less than 0.03 s), as expected, due to low mass flow rate through out the leak hole. Reduction in the pressure drop is greater with the increase of the water holdup in the mixture. After the flow reaches a new steady state (t > 0.025 s), the pressure drop reach the same level before the leakage. This level is greater for the cases where the

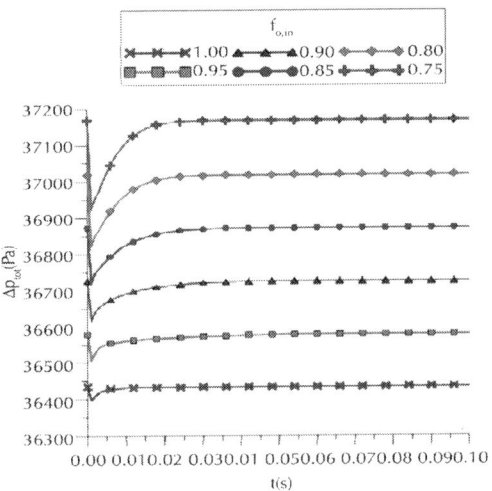

Figure 9: Pressure drop as a function of the time, for different oil volume fractions in the mixture at the pipe inlet ($U_{m,in}$ = 1.00 m/s).

water volume fraction in the mixture is greater, since the water has a density higher than the oil, increasing the static pressure drop portion, which is the most representative portion of the total pressure drop.

Figure 10 shows pressure drop as a function of the time, for different mixture velocities at the inlet, for a pipe section of 4 m length, where the leakage is located in the midpoint of this section. Similarly, it is visible the reduction in the pressure drop in the initial instants of the leakage. The transient period is too short, being less than 0.025 s. The higher pressure drop in transient state is obtained as the mixture velocity at the inlet is increased.

Figures 11 and 12 show, respectively, the total pressure and the oil velocity fields near the leakage section (in the transversal plan), for a case with the mixture velocity in the inlet 1.00 m/s and the oil and water volume fractions are 0.85 and 0.15, respectively. We can see that the leakage region is one zone of low pressure and high velocity. This behavior was found in all cases analyzed in this paper.

CONCLUSIONS

In this paper the hydrodynamic of two-phase flow in vertical pipe with a leakage is discussed. The study is related to heavy oil-water flow in the turbulent regime by using the ANSYS-CFX® 11.0 commercial software.

The simulations revealed the difficulty of detecting small leaks (less than 1% of the total mass flow rate transported in the pipeline), since the time interval of pressure drop observed is very short (less than 0.03 s). This research revealed too that in heavy oil-water flow it is more easy to detect leaks when the pipe operate with high fluid velocities and greater water volume fraction in the mixture at the pipe inlet. It was verified that the oil volume fraction in the mixture affect strongly the spread and the exit direction of the oil in the leakage hole.

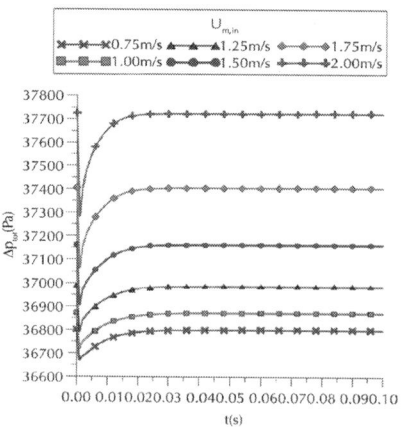

Figure 10: Pressure drop as a function of the time, for different mixture velocities at the pipe inlet ($f_{o,in} = 0.85$).

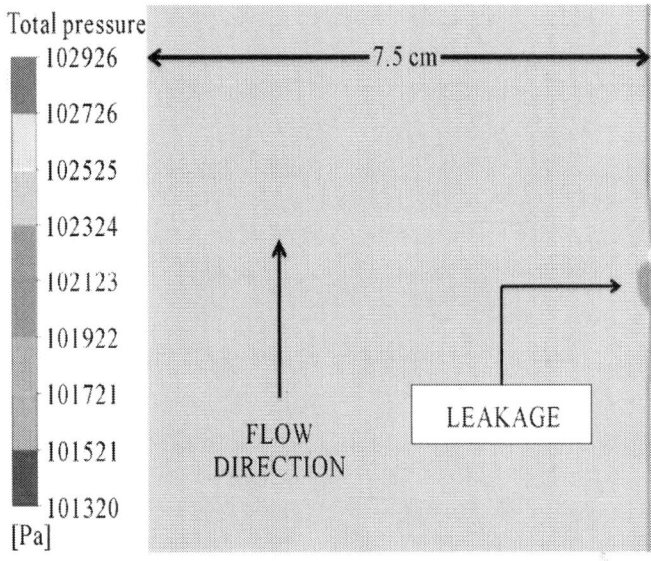

Figure 11: Pressure field near the leakage.

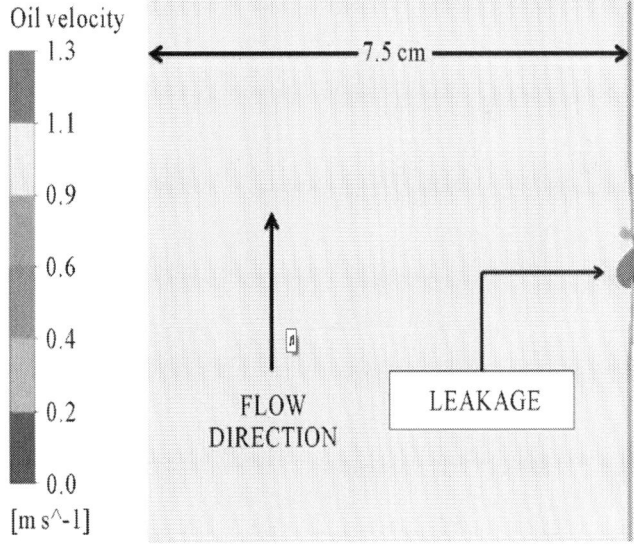

Figure 12: Oil velocity field near the leakage.

ACKNOWLEDGEMENTS

The authors would like to express their thanks to Brazilian researches agencies ANP/UFCG-PRH-25, CNPq, CAPES, FINEP and PETROBRAS S/A for supporting this work, and are also grateful to the authors of the references in this paper, that helped in the improvement of quality.

REFERENCES

1.   J. Y. Xu, D. H. Li, J. Guo and Y. X. Wu, "Investigations of Phase Inversion and Frictional Pressure Gradients in Upward and Downward Oil—Water Flow in Vertical Pipes," International Journal of Multiphase Flow, Vol. 36, No. 11-12, 2010, pp. 930-939.doi:j.ijmultiphaseflow.2010.08.007

2.   S. I. Kam, "Mechanistic Modeling of Pipeline Leak Detection at Fixed Inlet Rate," Journal of Petroleum Science and Engineering, Vol. 70, No. 3-4, 2010, pp. 145-156.doi:j.petrol.2009.09.008

3.   X. H. Lu, Y. J. Sang, J. Z. Zhang and Y. Y. Fan, "A Pipeline Leakage Detection Technology Based on Wavelet Transform Theory," Proceedings of IEEE Annual International Conference on Information Acquisition, Shandong, 20-23 August 2006, pp. 1432-1437.doi:10.1109/ICIA.2006.305966

4.   R. Hu, H. Ye, G. Wang and C. Lu, "Leak Detection in Pipelines Based on PCA," Proceedings of 8th International Conference on Control, Automation, Robotics and Vision, Kunming, 6-8 December 2004, pp. 1985-1989. doi:10.1109/ICARCV.2004.1469466

5.   H. V. Silva, C. K. Morooka, I. R. Guilherme, T. C. Fonseca and J. R. P. Mendes, "Leak Detection in Petroleum Pipelines Using a Fuzzy System," Journal of Petroleum Science and Engineering, Vol. 49, No. 3-4, 2005, pp. 223-238. doi:j.petrol.2005.05.004

6.   L. Dong, S. Chai and B. Zhang, "Leak Detection and Localization of Gas Pipeline System Based on Wavelet Analysis," Proceedings of 2nd IEEE International Conference on Intelligent, Control and Information Processing, Harbin, 25-28 July 2011, pp. 478-483. doi:10.1109/ICICIP.2011.6008290

7.   C. Verde, "Multi-Leak Detection and Isolation in Fluid Pipelines," Control Engineering Practice, Vol. 9, No. 6, 2001, pp. 673-682. doi:10.1016/S0967-0661(01)00026-0

8.   C. G. Mothé and C. Silva, "Heavy-oil—Reserves and World Production," TN Petróleo, Vol. 57, 2007, pp. 76-81. http://www. tnpetroleo.com.br/download.php/revista/download/i/33/nome/ TN57_Artigos.pdf

9.   ANSYS, "CFX-Solver Theory Guide (Release 11.0)," ANSYS, Inc., Canonsburg, 2006.

Chapter 2

Generation of Arbitrary Pressure Pulsation of Wide Frequency Range for Flow Meter Testing in a Laminar Gas Pipeline

Mitsuhiro Nakao[1], Tomonori Kato[2], Takashi Oowaku[3], Hirohisa Sakuma[3], Toshiharu Kagawa[4]

[1]Department of Mechanical Engineering, Kagoshima University, Kagoshima, Japan

[2]Faculty of Engineering, Fukuoka Institute of Technology, Fukuoka, Japan

[3]Tokyo Gas Co., Ltd., Yokohama, Japan

[4]Precision and Intelligence Laboratory, Tokyo Institute of Technology, Yokohama, Japan

ABSTRACT

The present paper presents a device to test flow meters under an arbitrary pressure pulsation in a gas pipeline with a laminar flow containing frequency components up to 50 Hz, with the amplitude reaching hundreds of pascals. In order to reduce flow noise, the device has a strainer-like element connected to a pipeline under test and uses an open-loop control law based on the frequency response test. The control signal is calculated by adding the inputs to obtain each of the sinusoidal waves included in the original wave, which was decomposed by Fourier analysis. The validity of the developed method is demonstrated through the generation tests of superimposed pressure waves containing frequency components up to 50 Hz. Analysis of the relative uncertainty demonstrated the relative uncertainty to be less than 10% when the generated pressure is larger than 360 Pa.

INTRODUCTION

Pulsations are a major source of error in flow measurements, as pointed by a number of researchers[1] -[3] . The generated error depends on both the type of the flow meter and the piping system used [1] -[3] . It is often the case that pipe flows are turbulent. In an actual pipeline having branches, however, there sometimes exist laminar flows provoked by unsteady flows in other branches. In a pulsating laminar flow, a periodically changing velocity distribution causes a relatively large error. In order to understand the performance of a flow meter under these circumstances, we need to examine a flow meter installed in a pipeline or duplicate the original pulsation in a laboratory experiment. The former approach is difficult because the true value of the flow rate is unknown. A laboratory-size instrument that can reproduce the original pulsation is needed for ease of testing.

A number of researchers have investigated unsteady flow rate testing devices, which can be categorized into flow rate control devices [4] -[6] and pressure control devices [7] . As an example of the former type, Durst et al. [4] proposed a mass flow controller consisting of a valve and a laminar element controlled by an open-loop control. This device is designed to work up to 125 Hz, and results up to 25 Hz are demonstrated. However, a disadvantage of the device is that the temperature of the gas discharged from the device changes during testing. Kawashima et al. [5] proposed an unsteady flow rate generator (UFG) using a chamber referred to as the isothermal chamber. The UFG has an advantage in that gas discharged from the UFG has an approximately constant temperature and a disadvantage in that the generation time is limited. Funaki et al. [6] modified the UFG to generate unsteady flows continuously. The modified UFG works well when flow rate to be examined is known such as in a frequency response test. However, it is difficult to evaluate the characteristics of a flow meter in a practical situation because the actual unsteady flow rate is rarely measureable. The pressure control device was proposed by Kato et al. [7] . The pressure control device controls the pressure pulsation at the inlet of a pipeline in order to generate an unsteady flow in order to evaluate the characteristics of a flow meter. An arbitrary pressure pulsation generating device (APPD) developed by Kato et al. [7] uses a feedback law based on the state equation of gas obtained by assuming that the pipeline under test can be treated as a lumped parameter system, and can reproduce pressure pulsations with small noise. This method can only reproduce a superimposed pressure wave including frequency components of up to 5 Hz. However, the actual frequency involved in the actual pulsating flow sometimes reaches tens of hertz. If the upper limit frequency of the APPD is improved, most actual pressure pulsations can be reproduced.

The objective of the present study is to develop an arbitrary pressure pulsation generating device, which can generate arbitrary pressure pulsations containing frequency components of up to 50 Hz, with an amplitude reaching hundreds of pascals. Such pressure pulsation occurs in real city gas pipelines, in which the line pressure is approximately 2.3 kPa, and a gas governor installed in the line has a natural frequency of 5 Hz or lower. It is usually satisfiable that the upper limit frequency is ten times of the natural frequency, but has not been achieved in the APPD. The present device has a configuration that is similar to that of an APPD, while using a different control law and concept. As the control law, an open-loop control law based on the frequency response characteristics is used to achieve a higher applicable frequency range. Instead of the large quasi-isothermal chamber used in a previous study [7] , a strainer-like element, such as a small quasi-isothermal chamber, is introduced in order to reduce the flow noise.

EXPERIMENTAL APPARATUS

Generation Principle

Pressure pulsations arise from various sources, such as the switching operation of solenoid valves in gas pipelines for domestic use, and are distributed in space and time. The duplication of the original pressure pulsations is practically impossible because an actual gas pipeline with branches, which is longer by tens of meters, is not practical for reproduction in an experimental environment.

In the present study, a simulated version of the gas pipeline focused only around a flow meter is constructed in order to scale down the length direction. If a variable, such as pressure, on the boundary condition of the pipeline is adjusted to be more or less the same state, as compared to the corresponding location in the original pipeline, the simulated pipeline would have a pressure pulsation that is more or less correct, as compared with the original pipeline (See Figure 1). Although the boundary conditions at both ends of the simulated pipeline should be ideally controlled, these boundary conditions complicate the control law. The pipeline downstream of the flow meter can be constructed in an experimental environment, because this pipeline is only a few meters in length. Thus, the present paper describes results for the case in which only the upstream side of the pipeline is controlled. This method is valid when the controlled location is not on a node of a major frequency component included in the original wave.

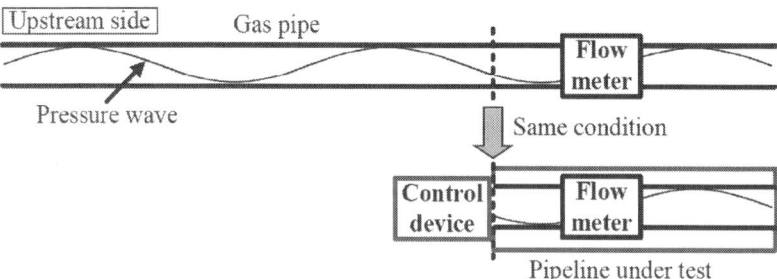

Figure 1: Generation principle of pressure pulsation.

Experimental Configuration

The experimental configuration of the present device is shown in Figure 2. Air, supplied by an air compressor, passes through a buffer tank with a volume of 2.6×10^{-2} m³. Then, the air pressure is reduced to approximately 0.6 MPaG, and the air flows into quasi-isothermal chamber 1. The flow rate, flowing from quasi-isothermal chamber 1, into a strainer-like element via a servo valve (MPYE-5-1/8-LF-010B, FESTO Co., Ltd.), is changed by controlling the input voltage to the servo valve. As a strainer-like element, a quasi-isothermal chamber, crafted by tightly stuffing copper wool into an equal tee fitting of type 1, having an outer diameter of 110 mm, implemented in relation to JIS standard [8] , is used. A 4-m-long steel pipe, with an internal diameter of 28 mm is used as the piping system under test. Pressure P_0 was measured using a semiconductor-type pressure sensor (XT190M-100A, Kulite Co., Ltd.), and pressures P_1 and P_2 were measured using piezoresistive differential pressure sensors (5 INCH-D-4V, All Sensors Co., Ltd.). All measured values were recorded using a computer, via an A/D convertor with a sampling frequency of 5 kHz. A D/A convertor gave the input voltage of the servo valve with the same frequency.

FREQUENCY RESPONSE TEST

First, the sonic conductance of the servo valve was obtained by experiment, using another experimental setup, as shown in Figure 3. Only the inlet port and one outlet port of the servo valve were used, and the other ports were closed with clinchers. The upstream pressure of the servo valve was set to approximately 0.6 MPaG in order to maintain a choked flow condition during the experiment. The volumetric flow rate passing through the servo valve was measured using a dry gas meter (DC-5A, SHINAGAWA, Co., Ltd.). During the test, the air temperature was 293 K, and the atmospheric pressure was 102 kPa. The volumetric flow rate passing through the servo valve in the chocked flow condition is expressed as follows [9] :

$$Q = CP_0\sqrt{\frac{293}{\theta}}$$

(1)

where C = sonic conductance, dm³/(s·MPa); P_0 = upstream pressure, MPa (G); Q = volumetric flow rate, dm³/s (ANR); = air temperature, K. The sonic conductance of the servo valve was calculated by substituting the measured values into Equation (1). Figure 4 shows the obtained characteristics of the input voltage E and the sonic conductance C and an approximated curve calculated by the method of least squares for 0 ≤ E ≤ 4.5 V, where C is given in terms of E as follows:

$$C = f(E) = -1.775 \times 10^{-1} E^2 - 8.715 \times 10^{-1} E + 7.627 \tag{2}$$

The quasi-sinusoidal flow rate is easily pumped into quasi-isothermal chamber 2 when the input voltage is given as follows:

$$E = E_{ave} + E_{amp} \sin(2\pi f t) \tag{3}$$

because the sonic conductance has strong linearity against the input voltage.

An example of the pressure response when the input voltage is set to E = 3.0 + 1.0sin (2ϖ20t) is illustrated in Figure 5. Pressure P_2 at the pipe outlet changed within 0.1 s of the start of the experiment and was approximately constant at 0 thereafter. Pressure P_1 largely pulsated due to the stepwise change in the flow rate within 0.4 s of the start of the experiment. After 1.5 s, a quasi-sinusoidal pressure wave arose. This result implies that the inlet flow rate to isothermal chamber 2 and the pressure response of P_2 have a linear relationship. In other words, when a sinusoidal flow rate with a frequency is input, this device will produce a sinusoidal pressure wave including only the input frequency. In order to examine this supposition, frequency response tests were performed on the device. In particular, the pressure response of P_1 was measured when the input voltage was given by Equation (3), in which E_{ave} was 3.0 V, E_{amp} was 0.1, 0.5, and 1.0 V, and f was a frequency selected from among 1 Hz to 50 Hz frequency range with 1-Hz increments.

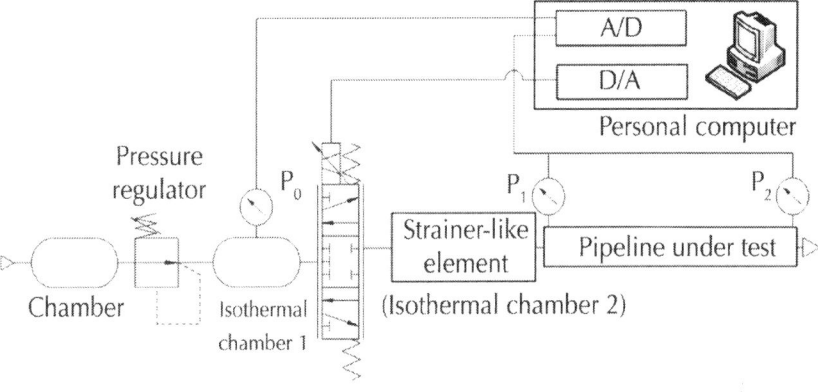

Figure 2: Pneumatic circuit of the developed equipment.

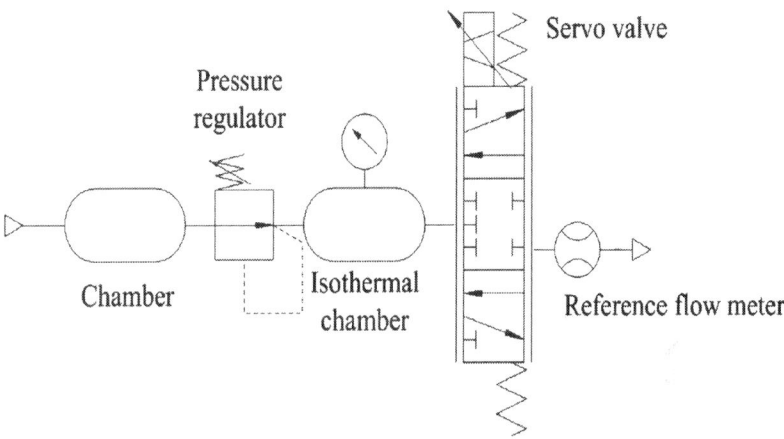

Figure 3: Experimental apparatus used in the sonic conductance test.

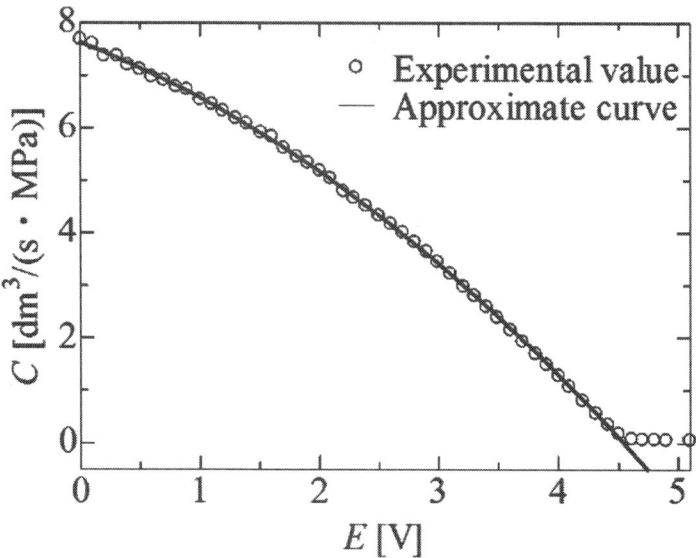

Figure 4: E-C characteristics of the servo valve.

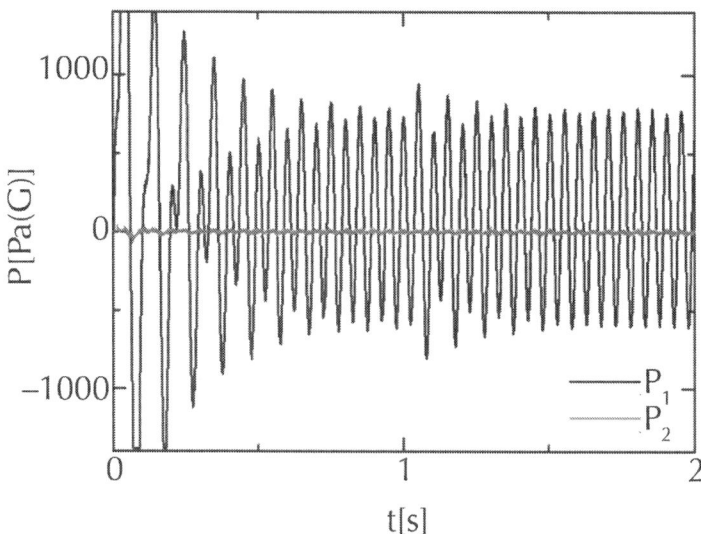

Figure 5: Measured pressures with an applied voltage of 3.0 + 1.0sin (2ϖ20t).

Each experiment was started at 0 s, and the data recording was started at 2 s and stopped at 4 s, for a total data recording time of 2 seconds. The data were then resolved into frequency components using the mixed radix fast Fourier transform algorithm, and the corresponding amplitudes and phases were obtained. Figure 6 represents a pressure response wave form of P_1, in the case of E_{amp} = 0.1 and 1.0 V, and f = 1.0 Hz, and Figure 7 shows the corresponding frequency components. Figure 8 and Figure 9 show similar results for the case of E_{amp} = 0.1 V and f = 10 Hz, and Figure 10 and Figure 11 show the results for case of E_{amp} = 0.1 and 1.0 V and f = 30 Hz. The results for the case in which E_{amp} = 0.1 V in Figure 7, except for the 0-Hz component, the 1-Hz component of a frequency similar to the given frequency of pulsation dominates, whe- reas the components around 10 Hz, corresponding to the Helmholtz resonance frequency, also have similar amplitude levels. On the other hand, a comparison of the results for E_{amp} = 0.1 V and 1.0 V reveals that the component at 1 Hz has a much larger amplitude for the case in which E_{amp} = 1.0 V, whereas the components around 10 Hz have approximately equal amplitudes in both cases. This is because the isothermal chamber acts as a porous material, which enhances small turbulent noise[10] while reducing large turbulent noise, and allows passage of the pressure wave [11] . The quasi-isothermal chamber is filled with very tightly packed copper wool [5] . Therefore, fluid flows through the narrow gaps in the wool, in the same manner as a porous material. The pressure form illustrated in Figure 8, is an almost pure sinusoidal waveform, because the given frequency is the natural frequency of the system. Actually, Figure 9 reveals that the frequency component equivalent to the natural frequency has an inordinate amplitude compared to the other components. Figure 10 and Figure 11 show similar results for a frequency of 30 Hz. In all of the cases presented above, the sinusoidal pressure responses with respect to sinusoidal inlet flow rates were obtained when the amplitude is relatively large.

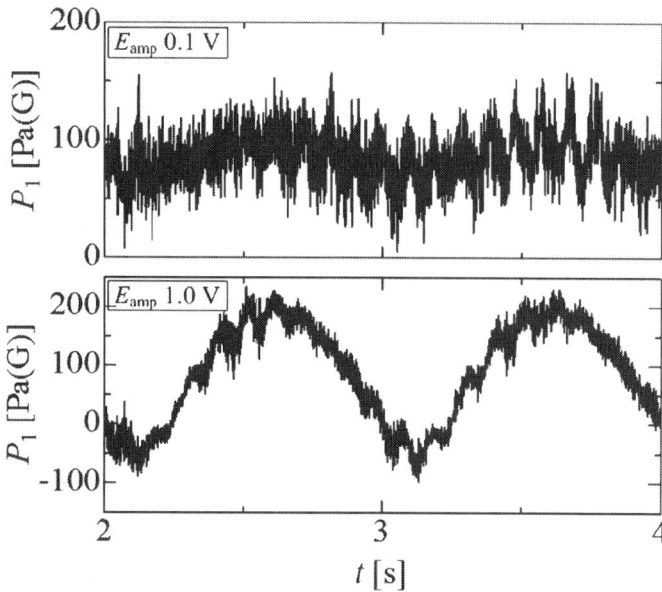

Figure 6: Measured pressures with the applied voltage of (upper) 3.0 + 0.1sin (2ϖt) and (lower) 3.0 + 1.0sin (2ϖt).

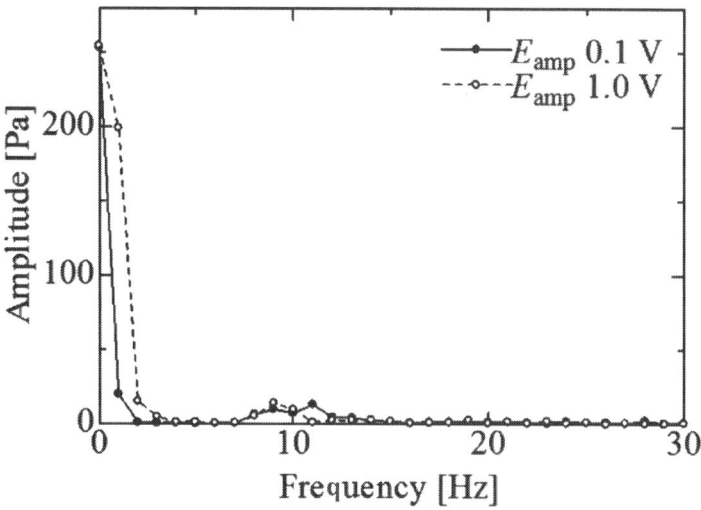

Figure 7: Frequency components of the wave forms illustrated in Figure 6.

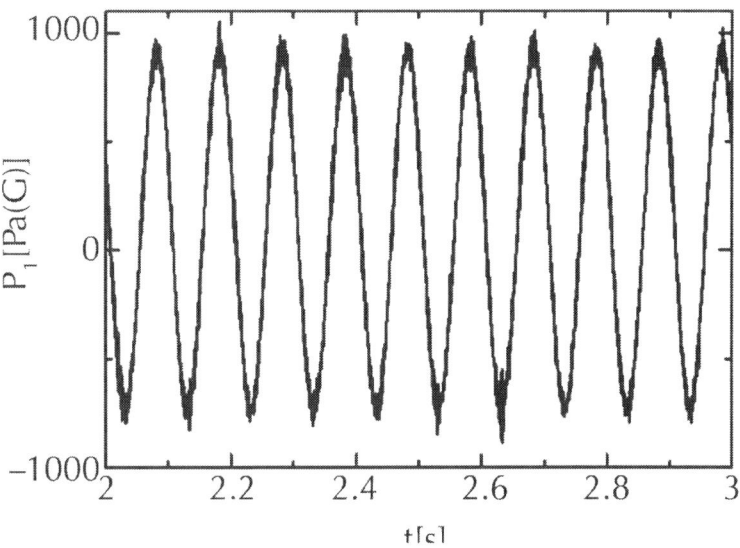

Figure 8: Measured pressures for an applied voltage of 3.0 + 0.1sin (2ϖ10t).

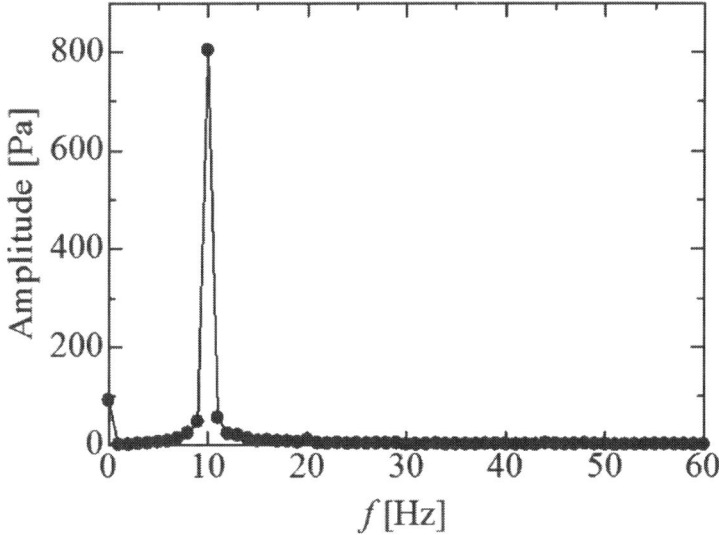

Figure 9: Frequency components of the wave form illustrated in Figure 8.

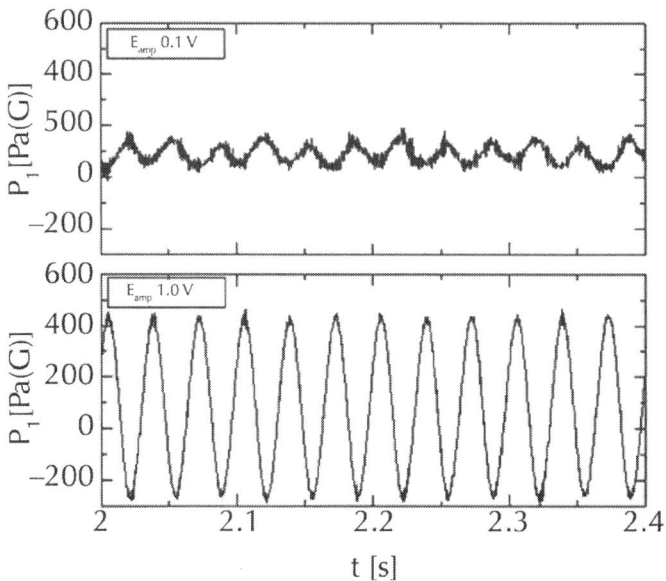

Figure 10: Measured pressures for applied voltages of (upper) 3.0 + 0.1sin (2ϖ30t) and (lower) 3.0 + 1.0sin (2ϖ30t).

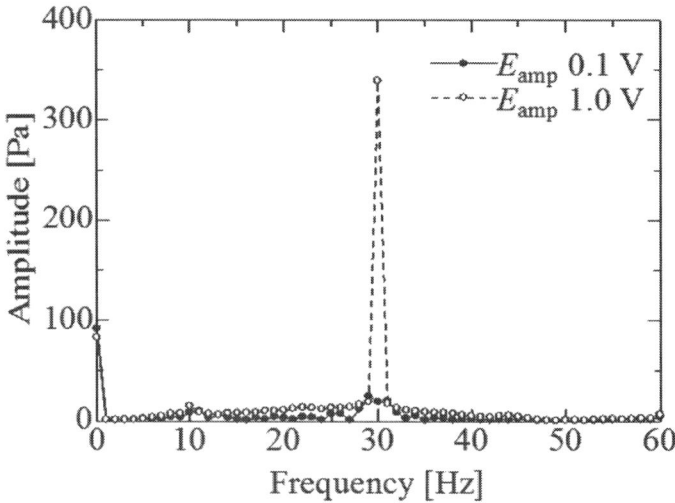

Figure 11: Frequency components of the wave forms illustrated in Figure 10.

The pressure response of a UFG was also investigated. The difference between a UFG and the present device is the existence of the strainer-like element. A UFG was constructed by removing the strainer-like element from the experimental apparatus shown in Figure 2. The pressure responses are summarized in Figure 12 for the case in which the input voltage is given by Equation (3), where E_{ave} was 3.0 V, E_{amp} was selected so as to obtain a pressure amplitude of up to hundreds of pascals, and f was 1.0, 10, or 30 Hz. Comparison of Figure 6, Figure 8, Figure 10, and Figure 12indicate that pressure wave forms generated with a UFG have a much larger turbulent noise compared to those generated by the present method. Furthermore, pressure wave forms generated with a UFG are non-linear with respect to the input flow rate. This result demonstrates the effectiveness of the strainer-like element.

METHOD OF GENERATING AN ARBITRARY PULSATING FLOW

Open-Loop Control Law

Among the most important facts presented herein is that the superposition of waves is valid, because the pressure wave addressed in this problem is a small-amplitude pressure wave that has a linear relation between the flow rate and the pressure pulsation. This means that if we can successfully generate individual pressure waves with a single frequency component included in an arbitrary pressure wave, we can reproduce an arbitrary pressure wave by superimposing the individual pressure waves.

Next, we present a definitive procedure for reproducing a target pressure pulsation. In practice, most periodic functions can be expanded in a Fourier series and can be described as a sum of trigonometric functions. The target pressure pulsation is generally expressed as follows:

$$P = \sum A_i \sin\left(2\pi f_i t + B_i\right)$$

(4)

Here, we consider the following single frequency component:

$$P_i = A_i \sin\left(2\pi f_i t + B_i\right)$$

(5)

When the pressure of sine wave pulsation given in Equation (5) is reproduced, the inlet flow rate Q satisfies the following relation:

$$Q_i = A_{Gi} \sin\left(2\pi f_i t + B_{Gi}\right)$$

(6)

Next, the following relations are determined experimentally in advance:

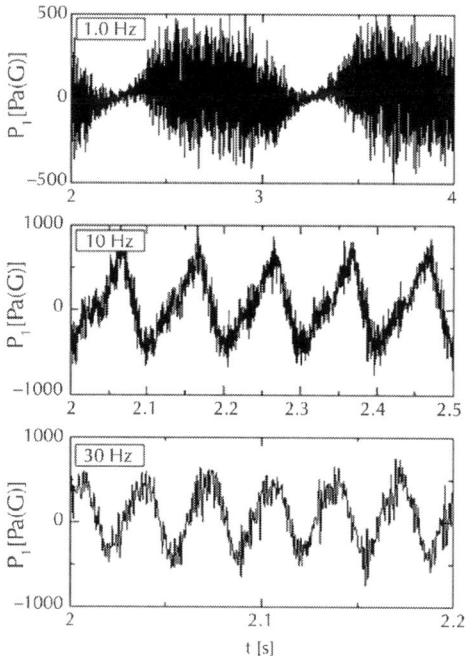

Figure 12: Pressure pulsations generated by a UFG with respect to three sinusoidal inputs: (upper) 1.0 Hz, (middle) 10 Hz, and (bottom) 30 Hz.

$$\frac{A_i}{A_{Gi}} = \text{Gain}_i$$

(7)

$$B_i - B_{Gi} = \phi_i \tag{8}$$

Then, Equation (6) can be transformed as follows:

$$Q_i = \frac{A_i}{\text{Gain}_i} \sin\left(2\pi f_i t + B_i - \phi_i\right) \tag{9}$$

Considering the choked flow condition, the transformation of Equation (9) yields the following equation:

$$C_i = \frac{A_i}{P_0 \cdot \text{Gain}_i} \sqrt{\frac{\theta}{293}} \sin\left(2\pi f_i t + B_i - \phi_i\right) \tag{10}$$

When we input the voltage corresponding to the sonic conductance given by Equation (10) to the servo valve, the pressure of sine wave pulsation given by Equation (5) is generated. If the target pressure pulsation has a number of frequency components, we can calculate the corresponding sonic conductance for each individual frequency by repeating the above-described process. The sonic conductance needed to reproduce the target pressure pulsation is then calculated as follows by taking the sum of the components based on the superposition of waves:

$$C = \sum C_i \tag{11}$$

Finally, the input voltage is calculated as follows using the inverse relation between the input voltage and the sonic conductance of the servo valve:

$$E = f^{-1}\left(C\right) \tag{12}$$

Results of Pressure Pulsation Generation

The authors examined pulsating flow generation tests using the experimental equipment shown inFigure 2 in order to check the validity of the proposed method. Generation tests involving three sinusoidal inputs having frequencies of 1.0, 20, or 50 Hz were first conducted. Using the above-described procedure, the input voltages were obtained by open-loop control, where the reference voltage was calculated using a computer and a measured pressure of P_0. Figure 13 shows the generation results and indicates that all of the generated pressures matched the reference values very well with no feedback compensation. Figure 14 shows the frequency components included in the reference signal. The wave forms shown in Figure 13 are decomposed by FFT. Comparison of these figures shows good agreement. We also examined a frequency response test of the APPD [7] against pressure pulsations with amplitude of 100 Pa. The result is summarized as a bode diagram shown in Figure 15. It is demonstrated that the APPD could not reproduce pressure waves with a frequency higher than 5 Hz since the phase delay is 135 degree at 5 Hz. Thus, the proposed open-loop control method can reproduce higher-frequency waves, as compared to the previous method, with a relatively large amplitude, which was difficult using the previous method.

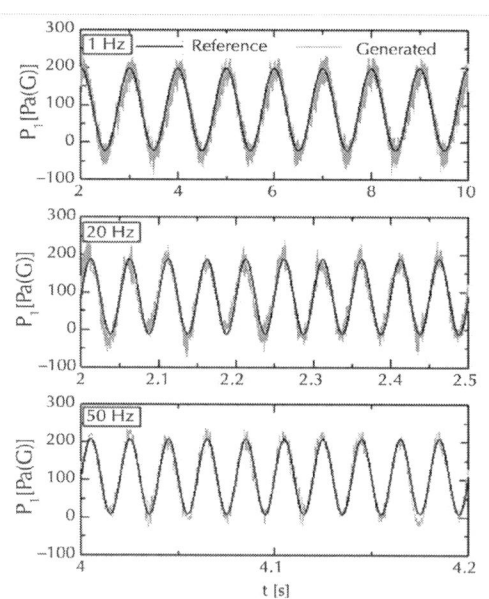

Figure 13: Pressure pulsations generated by the present method with respect to three sinusoidal inputs: (upper) 1.0 Hz, (middle) 20 Hz, and (bottom) 50 Hz.

A superimposed pressure wave including three frequency components of 5, 20, and 50 Hz was also generated. The results are shown in Figure 16. The generated super-imposed pressure pulsation shows excellent agreement with the reference. Figure 17 shows the frequency components included in the reference signal and the wave forms illustrated in Figure 16 decomposed by FFT. The frequency spectrum of the generated pulsation has a small error in comparison with the reference values, which may indicate that the pressure wave occurring in the present device is not a linear wave, but can be approximately treated as a linear wave.

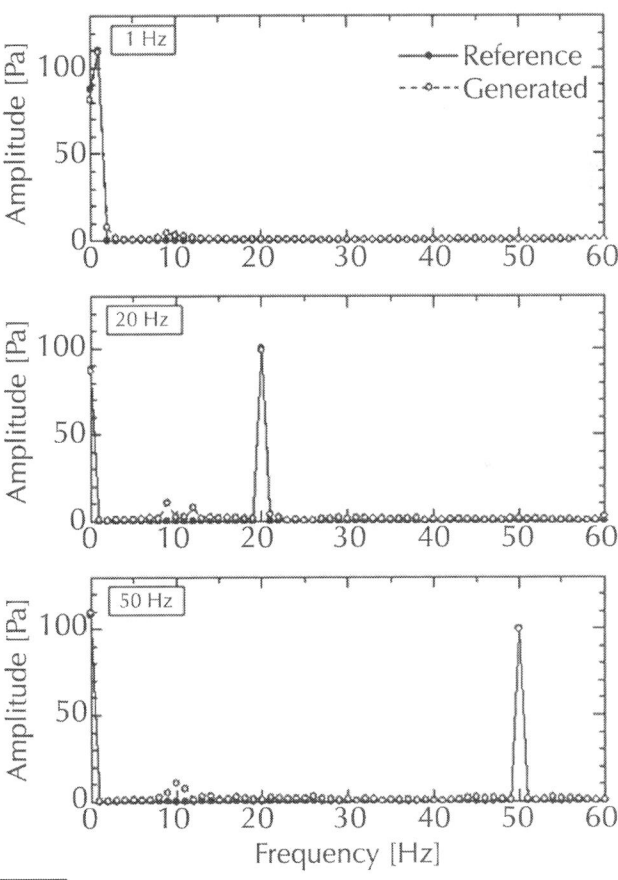

Figure 14: Frequency components of the wave forms illustrated in Figure 13: (upper) 1.0 Hz, (middle) 20 Hz, and (bottom) 50 Hz.

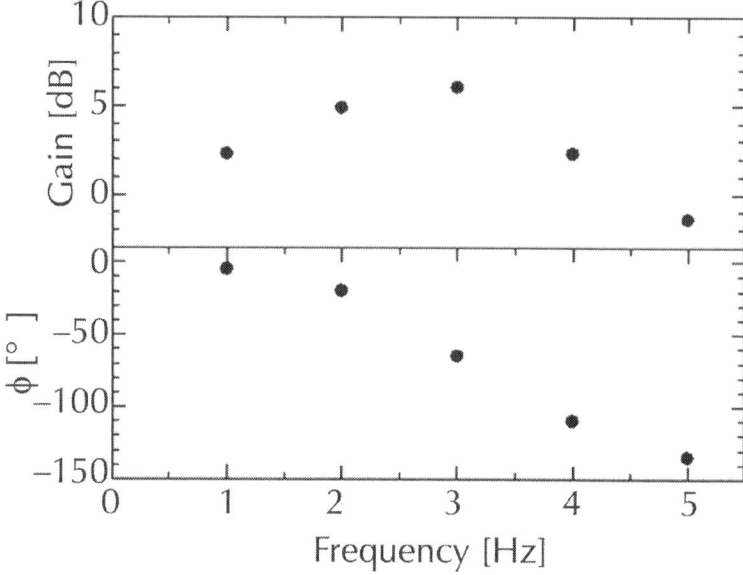

Figure 15: Bode diagram of the APPD.

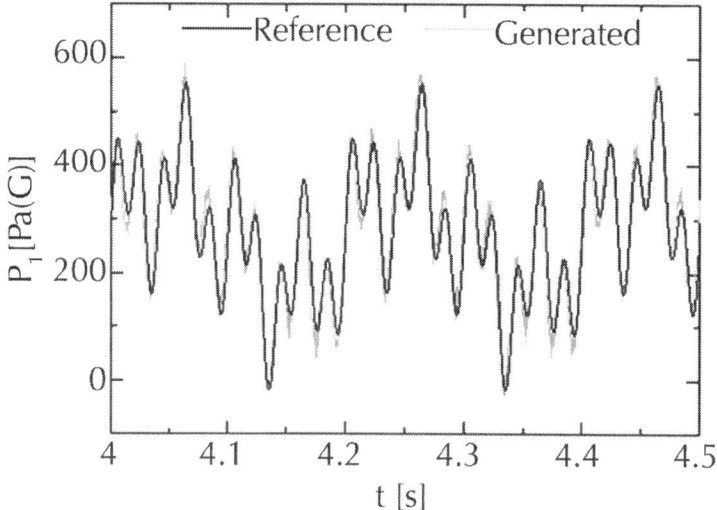

Figure 16: Generation results of the superimposed pressure wave.

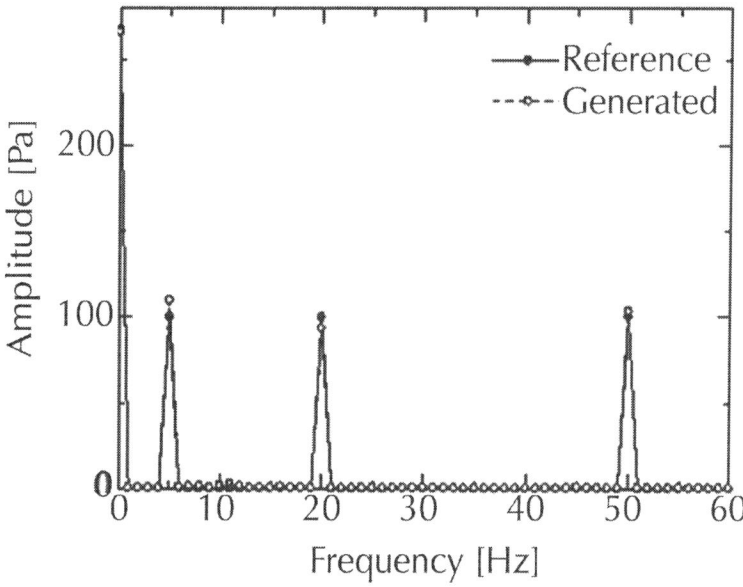

Figure 17. Frequency components of the wave form illustrated in Figure 15.

There are three factors that cause the non-linearity: hysteresis of the servo valve, measurement noise, and induced noise downstream of the servo valve. The hysteresis generates causes non-linearity, and the noises enhance frequency components around the natural frequency, which is not in targeted signal. The validity of the present device as an approximative device for generating arbitrary pressure pulsations is thus confirmed.

UNCERTAINTY ANALYSIS

The present device generates pressure pulsations by means of open-loop control of the inlet flow rate into the test pipeline based on known frequency characteristics. Errors are generated in the determination of the frequency characteristics, the inlet flow rate generation, and the noise caused by the servo valve. The relative uncertainty analysis of the present device is discussed below [12] .

The errors related to the frequency characteristics are associated with the inlet flow rate and the wall pressure measurement. The uncertainty of the inlet flow rate depends on the temperature, the measurement of the upstream pressure, and the sonic conductance of the servo valve given by Equation (1). The uncertainty due to the change in the supply temperature is less than 1% when the supply temperature is larger than 250 K, because the temperature change in the isothermal chamber is only approximately 2.5 K [6] . The uncertainty due to the temperature measurement is 0.8 K. The uncertainty of the measurement of the upstream pressure is 0.8%. The uncertainty of the sonic conductance is estimated to be 1%. The standard uncertainty of the inlet flow rate is given by following formula from the propagation of uncertainty:

$$\frac{\delta Q}{Q} = \sqrt{\left(\frac{\delta C}{C}\right)^2 + \left(\frac{\delta P_0}{P_0}\right)^2 + 0.25\left(\frac{\delta \theta}{\theta}\right)^2}$$

(13)

The uncertainty of the wall pressure measurement is 1.1%. Thus, the standard uncertainty of the frequency characteristics is given by following equation:

$$\frac{\delta Gain}{Gain} = \sqrt{\left(\frac{\delta P_1}{P_1}\right)^2 + \left(\frac{\delta Q}{Q}\right)^2}$$

(14)

When the temperature is larger than 250 K, which is applicable in nearly all practical cases, the uncertainty of the inlet flow rate is less than 1.4% from Equation (13), and the uncertainty of the frequency characteristics is less than 1.8% from Equation (14).

The noise and nonlinearity generate errors. The uncertainty of this generated pressure is given by following equation:

$$\delta P_{1,generated} = \sqrt{\frac{\sum \left(P_{1,reference} - P_{1,generated} \right)}{n-1}}$$

(15)

where n = a number of the sampled data. All data illustrated in Figure 13 and Figure 15 was used to calculate Equation (15). The uncertainty depends on frequency as summarized in Table1 The uncertainty is less than 35 Pa among all frequencies.

The total relative uncertainty of the present device is obtained by combining the above uncertainties as follows:

$$\frac{\delta P_1}{P_1} = \sqrt{\left(\frac{\delta Q}{Q}\right)^2 + \left(\frac{\delta Gain}{Gain}\right)^2 + \left(\frac{\delta P_{1,generated}}{P_{1,generated}}\right)^2}$$

(16)

Equation (17) has validity when the temperature is larger than 250 K:

$$\frac{\delta P_1}{P_1} \leq \sqrt{5.1 \times 10^{-4} + \frac{1225}{P_{1,generated}^2}}$$

(17)

For example, the total relative uncertainty of the present device is less than 10% when the generated pressure is greater than 360 Pa. The greater the generated amplitude, the smaller the uncertainty of the present device.

CONCLUSIONS

In the present paper, we have proposed a device that can generate arbitrary pressure pulsations containing frequency components of up to 50 Hz with an amplitude reaching hundreds of pascals. A small quasi-isothermal chamber was introduced to the device as a strainer-like element in order to reduce turbulent noise. Frequency response tests demonstrated that pulsations with amplitudes of hundreds of pascals occurring in the device are linear waves, thus verifying the validity of the superposition principle, and the open-loop control law was developed based on the superposition principle. The control signal is calculated by adding up inputs to obtain each sinusoidal wave included in the original wave, which is decomposed by Fourier analysis. Reproduction tests of single sinusoidal pressure waves were carried out, and the results confirmed that the developed method can reproduce pressure pulsations of up to 50 Hz with small turbulent disturbances, which is not possible using existing devices. Furthermore, the proposed device also generated superimposed pressure waves containing three frequency components. Analysis of the relative uncertainty of the proposed device revealed a relative uncertainty of less than 10% for the case in which the amplitude of the pressure pulsation is greater than 360 Pa. The proposed device successfully improved the applicable frequency range of the pressure-control-type flow meter test device. Using the proposed device and pressure measurement in an actual pipeline, users can test flow meters under a wider range of pressure pulsations. The quasi-linear wave assumption loses validity in the turbulent flow case. A more complex control method will be needed to compensate the non-linearity to extend the present device to the turbulent flow case.

Table 1: Frequency dependency of the $\delta P_{1, \text{generated}}$

Frequency	1 Hz	20 Hz	50 Hz	5 Hz + 20 Hz + 50 Hz
$P_{1, \text{generated}}$	20.9 Pa	23.4 Pa	18.7 Pa	34.6 Pa

ACKNOWLEDGEMENTS

The authors would like to thank Mr. T. Hanada and Mr. T. Chifu, former students of Kagoshima University, for their assistance in conducting the experiment.

REFERENCES

1.   Håkansson, E. and Delsing, J. (1994) Effects of Pulsating Flow on an Ultrasonic Gas Flowmeter. Flow Measurement and Instrumentation, 5, 93-101. http://dx.doi.org/10.1016/0955-5986(94)90042-6

2.   Hanson, B.R. and Schwankl, L.J. (1998) Error Analysis of Flowmeter Measurements. Journal of Irrigation and Drainage Engineering, 124, 248-256. http://dx.doi.org/10.1061/(ASCE)0733-9437(1998)124:5(248)

3.   Berrebi, J., Martinsson, P.E., Willatzen, M. and Delsing, J. (2004) Ultrasonic Flow Metering Errors Due to Pulsating Flow. Flow Measurement and Instrumentation, 15, 179-185. http://dx.doi.org/10.1016/j.flowmeasinst.2003.12.003

4. Durst, F., Heim, U., Ünsal, B. and Kullik, G. (2003) Mass Flow Rate Control System for Time Dependent Laminar and Turbulent Flow Investigations.Measurement Science and Technology, 14, 893-902. http://dx.doi.org/10.1088/0957-0233/14/7/301

5. Kawashima, K. and Kagawa, T. (2003) Unsteady Flow Generator for Gases Using Isothermal Chamber. Measurement, 33, 333-340. http://dx.doi.org/10.1016/S0263-2241(03)00003-4

6. Funaki, T., Kawashima, K., Yamazaki, S. and Kagawa, T. (2007) Generator of Variable Gas Flows Using an Isothermal Chamber. Measurement Science and Technology, 18, 835-842. http://dx.doi.org/10.1088/0957-0233/18/3/036

7. Kato, T., Oowaku, T., Sakuma, H. and Kagawa, T. (2012) Introduction of a Newly Developed Arbitrary Pressure Pulsation Generating Device for Evaluating the Characteristics of Gas Flow Meters and Sensors. JFPS International Journal of Fluid Power System, 5, 16-21.

8. JIS B2301 (1994) Screwed Type Malleable Cast Iron Pipe Fitting.

9. JIS B8390 (2000) Pneumatic Fluid Power Components Using Compressible Fluids Determination of Flow Rate Characteristics.

10. Takatsu, Y. and Masuoka, T. (2002) Production and Dissipation of Turbulence in Porous Media. Transactions of JSME series B, 68, 2601-2605.

11. Sueki, T., Takaishi, T., Ikeda, M. and Arai, N. (2010) Application of Porous Material to Reduce Aerodynamic Sound from Bluff Bodies. Fluid Dynamics Research, 42, 015004. http://dx.doi.org/10.1088/0169-5983/42/1/015004

12. ISO (1995) Guide to the Expression of Uncertainty in Measurement.

Chapter 3

Use of Corrosion Inhibitor in Solid Form to Prevent Internal Corrosion of Pipelines and Acidification Process

Fernando B. Mainier[1], and Pedro Ivo Canesso Guimarães[2]

[1]Escola de Engenharia, Universidade Federal Fluminense (UFF), Niterói, Rio de Janeiro, Brazil

[2]Instituto de Química, Universidade do Estado do Rio de Janeiro (UERJ), Rio de Janeiro, Brazil

ABSTRACT

Use of corrosion inhibitors in solid form promotes the development of a new technique for internal corrosion protection of oil & gas pipelines and operations of oil wells acidification, because the controlled dissolution of the corrosion inhibitor forms a surface on metallic parts, a protective film that prevents or minimizes undesirable reactions to corrosion. In addition, this technique has important social and environmental benefits, ensures the operator has a lower risk of contamination when handling the product, changes the type of industrial packing, facilitates transportation, reduces solvent use and consequently reduces the waste that normally results from the use of inhibitors. The purpose of this article is to present a class of solid corrosion inhibitor tested in the laboratory and offer proposals for its application in industrial pipes such as gas and oil pipelines.

INTRODUCTION

The study of the modification of polymers obtained by condensation or addition is of growing interest in several industrial segments because of the function of the reactive groups present in the polymers' structure. In general, this type of modification targets the synthesis of polymers with physico-chemical properties for industrial application as in the formulation of corrosion inhibitors. Obtaining the polymer: poly (2-vinyl-2-oxazoline), PEOX, in solid form is such an application. The solid characteristics of this substance make possible its use in acidification (stimulation) in oil wells and could result in benefits in terms of controlled dissolution and ease of packaging and transport packaging.

Corrosion inhibitors aim at the protection of metals and alloys, preventing or retarding corrosion reactions through formation of a monomolecular film-adsorbed surface exposed to corrosive medium [1] [2] .

The modified polymers in the form of solids can be used as corrosion inhibitors, but they must possess, inter alia, the following features: anti-corrosion protection capability, proper consistency to use controlled dissolution rate and chemical compatibility with the material, corrosive medium and other additives used in petroleum industry operations.

This work aims at presenting a corrosion inhibitor in solid form for use in the hydrochloric acid solutions commonly used in oil well stimulation and proposing solid inhibitor application techniques for use in segments of the petroleum industry such as oil and gas pipelines.

CHARACTERISTICS OF CORROSION INHIBITOR BASED ON THE POLY (2-VINIL-2-OXAZOLINE)

The polymer poly (2-vinyl-2-oxazoline) was obtained by the reaction of polyacrylonitrile with 2-amino-ethanol, in the presence of catalyst cadmium acetate, $[(Ac)_2Cd]$, under reflux to 150°C for 25 hours, as shown in the following reaction. After reflux, the reaction mixture was distilled to eliminate excess amino-ethanol, filtered, purified and dried under vacuum [3] [4]

This polymer has oxazolinic rings hanging on the main chain in place of nitrile groups and, consequently, the polymer thus formed will have the characteristics of a paraffin-like appearance associated with the properties of oxazoline-2-substituted. This substance was dark brown, solid, and elastic and its molecular weight was valued at 1000.

When a steel plate is immersed in a solution of hydrochloric acid the preferential reactions (anodic and cathodic) that occur on the metallic surface are:

$Fe - 2e^- \rightarrow Fe^{2+}$ (anodic reaction).

$2\,H^+ \rightarrow H_2 - 2e^-$ (cathodic reaction).

The Fe^{2+} ion formed by the attack of the acid on the carbon steel leaves the metal (Fe) for the solution, and consequently there is a migration of H^+ ions from the concentrated acid to the metal surface to form atomic hydrogen (H) and, soon afterwards, molecular hydrogen (H_2). The addition of an organic inhibitor-type system non-oxidizing acid reaction can lead to partial or even total with the H^+ ions dissociated in acidic solution, having spontaneously captured the positive charges by the inhibitor molecule, a process called protonation [2] [5] [6] .

Consequently there is intense competition between H^+ ions and protonated inhibitor molecules moving into areas where they can accumulate cathode electrons. Although the ionic mobility of the H^+ ions is much greater (because of the smaller size of the ion) than that of the protonated molecular inhibitor, there is an adsorption of the stable inhibitor on the metallic surface, which forms a barrier that prevents migration of H^+ ions to capture the electrons, thus preventing the formation of atomic hydrogen (H) and detachment of molecular hydrogen (H_2).

The adsorptive inhibitors or films are generally organic substances of high molecular weight that form a monomolecular film on the metal surface, preventing the development of electrochemical reactions.

Considering the protonation in acid medium the most likely mechanism is the adsorption of the oxazolinic rings on the metallic surface which forms a protective film and reduces or prevents the cathodic reactions.

Corrosion Laboratory Test and Results

In this paper, the corrosion coupons to represent the pipeline were made from an AISI 1020 steel sheet measuring $45 \times 15 \times 1.2$ mm. They were prepared with three different sandpapers: 150, 320 and 400, in that order. Then they were degreased with acetone, washed with deionized water and anhydrous alcohol and were finally dried with hot air and stored in desiccators for use in the tests. They were then weighed to the nearest 0.0001 g.

In laboratory experiments hydrochloric acid solution 10% (% mass) was used as a corrosive medium. The basic substances used as corrosion inhibitors were poly (2-vinyl-2-oxazoline) concentrations of 1000 mg/L, 2000 mg/L, 3000 mg/L and 4000 mg/L.

Gravimetric assays (weight loss) were performed in a glass vessel resistant to temperature variation with a capacity of 300 mL. The corrosion coupons were completely immersed in 150 mL of acid solution, leaving the remaining capacity of the vessel for the evolution of hydrogen (H_2) resulting from acid attack.

The glass vessel was kept at the correct temperature via a thermostatically controlled bath. The temperature was set at 60°C. The time for testing was 1 hour after exposure. Immediately after completion of the test, the corrosion coupons were removed from the corrosive medium, rinsed in water and alcohol and quickly dried in hot air, before being weighed with the same accuracy. The weight loss was determined according to ASTM G 31 - 72 [7]

The corrosion rate (CR) and the efficiency of corrosion inhibitors (E %) are defined by the following expression:

Corrosion rate = CR = $(W_o - W_i)/S.t$ (mg/cm²·h).

Efficiency = E % = 100 $(W_o - W_i)/W_o$.

Where:

W_o and W_i are the weight loss in the absence and presence of the inhibitor;

S = area (cm²);

t = exposure time, h.

The results of the laboratory tests carried out on four carbon steel coupons immersed in hydrochloric acid solution 10% (% mass) and corrosion inhibitor poly (2-vinyl-2-oxazoline) in concentrations of 1000 mg/L, 2000 mg/L, 3000 mg/L and 4000 mg/L are presented in Table 1 and the graph ofFigure 1.

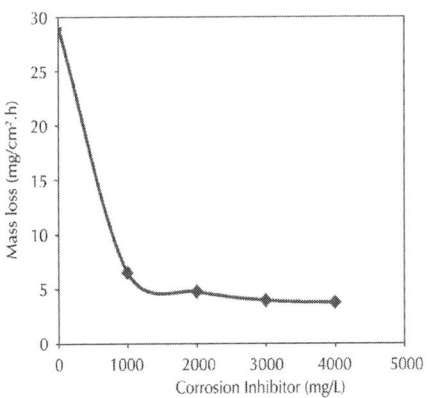

Figure 1: Corrosion of carbon steel in hydrochloric acid (10% mass) after additions of corrosion inhibitor at a temperature of 60°C and immersion for one hour.

Table 1: Corrosion rate of carbon steel in hydrochloric acid (10% mass) after additions of corrosion inhibitor at a temperature of 60°C and immersion for 1 hour

Concentrations: poly (2-vinyl-2-oxazoline, mg/L	Loss of mass (mg/ cm²·h)	Efficiency (%)
0	29.10	-------
1000	6.57	77.40
2000	4.80	83.50
3000	3.98	86.30
4000	3.80	87.90

APPLICATION OF CORROSION INHIBITORS IN OIL PIPELINES, GAS PIPELINES AND ACIDIFICATION SYSTEMS

The use of corrosion inhibitors in internal protection of pipes is a technique widely used in the production and transportation of oil and its derivatives. Applications are usually in liquid or emulsion form, and are organic and inorganic substances, or both.

The solvent has is physicochemistry characteristics of each inhibitor formulation. Solvents used for organic substances are toluene, glycols, alcohols (from methanol to hexanol) and kerosene. In the case of inorganic substances water is generally the solvent although alcohols are also used.

These formulations may also be associated with the demulsifying agent, antifoaming, dispersant, etc. The literature over the years has shown that most substances used are potentially toxic and the use of aromatics such as benzene has carcinogenic potential. Therefore less toxic substitutes are desirable [1] .

In the case of pipelines, gas pipelines and acidification systems the injection of corrosion inhibitors for internal protection is in liquid form with solvents such as the BTX family (benzene, toluene and xylene). The idea of using a solid corrosion inhibitor is to reduce or minimize the risk of contamination to humans (during packaging or operation) and the environment, as it requires no solvent.

Dissolution tests performed with the solid substance poly (2-vinyl-2-oxazoline) presented on the graph in Figure 1 show an order of efficiency of 87%.

Consequently the following proposals for the application of a corrosion inhibitor in solid form are made:

- Small spheres, pellets or soluble capsules injected into the pipe with special equipment;
- Compressed cylindrical rod, consumable and adapted equipment.

Small Spheres, Pellets or Soluble Capsules Injected with Special Equipment

Corrosion inhibitors can be manufactured in spherical or oval form or packaged in soluble capsules and injected by a gun into a gas pipeline, oil pipeline or acidification system. The mechanism drawn in Figure 2 shows that the small spheres dissolve in the flux to form a film adsorbed on the metallic surface parts for the necessary corrosion resistance.

Another way to use the inhibitor is to pack its solid particles in soluble capsules, as shown in Figure 3, in such a way that it will form a protective film on the metallic surface.

The injection system consists essentially of a pressurized tank attached to a valve and welded to a pipe flange, as shown in Figure 4. The pressurized reservoir containing the spheres or capsules of the inhibitor is controlled and programmed to release mode, so gradually the spheres or capsules adhere to the inside of the pipe.

Pressed Cylindrical Rod, Consumable and Adapted Equipment

This is a simple system, consisting essentially of a cylindrical rod which is introduced inside the pipe through a flange welded on the outside, as shown in Figure 5. The cylindrical rod dissolves and releases particles as a function of flow of the corrosive medium. The cylindrical rod is compressed continuously and controlled for the interior of the pipe as if wearing.

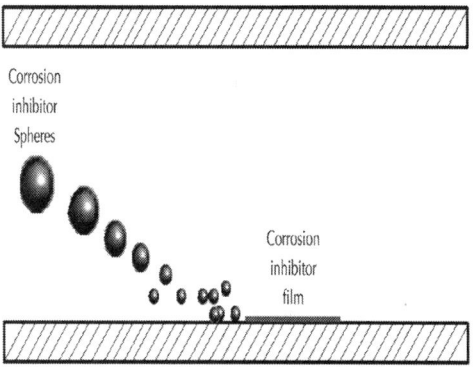

Figure 2: Mechanism of corrosion inhibitor.

Figure 3: Soluble capsules with solid corrosion inhibitor.

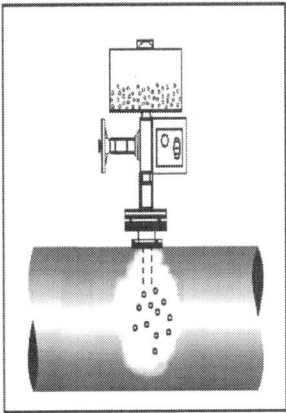

Figure 4: Injection system welded in pipe.

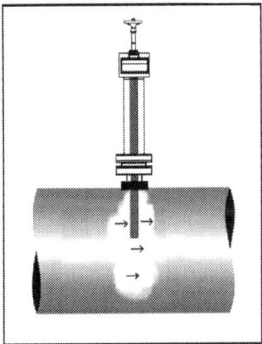

Figure 5: Corrosion inhibitor in the form of a cylindrical rod inside the pipe.

CONCLUSIONS

- Laboratory testing with hydrochloric acid solution showed that the inhibitor constituted of poly (2-vinyl-2- oxazoline) presented 87% efficiency in corrosion protection of carbon steel coupons. It is an excellent example of a corrosion inhibitor in solid form.

- The use of a corrosion inhibitor in solid form opens the door to new uses as there is no need for toxic organic solvents. This technology increases the operational safety of the worker handling the product and reduces environmental contamination.

- The use of a corrosion inhibitor in solid form as small spheres, packaged as a soluble capsule in the form of a compressed cylindrical rod, provides a new anti-corrosion technique that is easy to operate, particularly in gas and other pipelines and acidification systems.

REFERENCES

1. Fink, J.K. (2003) Oil Fields Chemicals. Gulf Professional Publishing, New York.

2. Mainier, F.B., Monteiro, L.P.C., Tavares, S.S.M., Leta, F.R. and Pardal, J.M. (2003) Evaluation of Titanium in Hydrochloric Acid Solutions Containing Corrosion Inhibitors. IOSR Journal of Mechanical and Civil Engineering, 10, 66-69.http://dx.doi.org/10.9790/1684-1016669

3. Monteiro, A.P., Guimarães, P.I.C. and Mainier, F.B. (1993) Synthesis and Application of Poly (Ethylene-2-oxazolina). Brazilian Congress of Polymers, 2, São Paulo, 5-8 October 1993 (in Portuguese)

4. Rodríguez, S., Abreu, A. and Cepero, A. (2003) Comparative Study of Three Oxazolines as Corrosion Inhibitors (Part II). CENIC Journal. Ciencias Químicas, 34, 15-20 (in Spanish).

5. Cruz, J., Martınez, R. and Genesca, J. (2004) Experimental and Theoretical Study of 1-(2-Ethylamino)-2-methylimidazoline as an Inhibitor of Carbon Steel Corrosion in Acid Media. Journal of Electroanalytical, 566, 111-121.

6. Bentiss, F., Trisnel, M. and Lagrenee, M. (2000) The Substituted 1,3,4-Oxadiazoles: A New Class of Corrosion Inhibitors of Mild Steel in Acidic Media. Corrosion Science, 42, 127-146. http://dx.doi.org/10.1016/S0010-938X(99)00049-9

7. ASTM G31-72 (1999) Standard Practice for Laboratory Immersion Corrosion Testing of Metals.

Hydrodynamics of Liquid Film in Helical Tubes

Mohammed Salah Hameed and Masab Kadhim
Jawad

Chemical Engineering Department, Higher Colleges of Technology, Sharjah, United Arab Emirates

ABSTRACT

Hydrodynamic experiments on a liquid film are carried out using water in both straight and helical tubes at angles of inclination ranging between 2.5° and 5° and on three different coil diameters (23.86 cm, 32.74 cm and 41.13 cm) for film Reynolds numbers ranging from 100 to 2000. The film thickness is measured by two micrometers, arranged to measure vertical and horizontal distances within the cross section of the tube. The results of film thickness are related to the hydraulic radius to characterize the film flow in both types of tube. Momentum transfer rates are shown to be higher in helical tubes than in the straight incline tube. An empirical correlation is presented for film thickness in the helical tube in terms of N_T (coil tube)/N_T (straight tube) for film Dean number ranging from 1 to 1000.

INTRODUCTION

The occurrence and applications of film flow in chemical engineering processes are numerous; among them are absorption, extraction, heat transfer, humidification and distillation. The Wetted-wall column, a well know film flow technique, is frequently used for the experimental determination of mass transfer coefficients.

The presence of curves or bends is unavoidable in the design of open channel [1] thus producing spiral current and cross-waves in addition to the unique features of super elevation produced by centrifugal force.

Throughout the last century research work was mainly focused on film flow over flat surfaces or channels and over vertical tubes. Since 1960, extensive literature has been published dealing with wavy gas-liquid interfaces and concentrating mainly on the conditions under which waves exist and their effect on the processes of heat, mass and momentum transfer.

Hopf (cited in [2]) conducted experiments in a rectangular channel of slope ranging from 0.5° to 3.5° and noted the influence of turbulence between the critical film Reynolds numbers ($\mathrm{Re}_{i,\ crit}$) ranging from 250 to 300. He found that wall roughness has no effect on $\mathrm{Re}_{i,crit}$ except for the smallest depths. Nusselt (cited in [3]) gave a theoretical treatment for smooth, laminar and two dimensional film flows and stated that:

$$\delta = \left(\frac{3 \upsilon^2}{g \sin \psi} \right)^{1/3} \left(\mathrm{Re}_f \right)^{1/3}$$

(1)

$$U_{av} = \frac{g \sin \psi}{3 \upsilon} \delta^2$$

(2)

$$\frac{U_s}{U_{av}} = 1.5.$$

(3)

$$f = \frac{6}{\mathrm{Re}_f}$$

(4)

$$N_T = \left(3\,\mathrm{Re}_f\right)^{1/3}$$

(5)

Jefferys [4] conducted experiments in channels at small slopes for a large range of Re_i. He confirmed the applicability of Equation (3) for the laminar region while the ratio of velocities decreases to 1.06 for the turbulent region. Cooper and Willey [5] determined experimental data for dilute sulfuric acid inside vertical tube and together with other workers found that up to Re_i equal to 350, the data are in excellent agreement with Equation (3). While Kirkbride's data [6] on film flowing outside a vertical tube deviated positively from the theoretical film thickness.

Fiend (cited in [2]) gave an experimental correlation for film flow of water and aqueous solutions with counter-current air in vertical tubes in turbulent region ($\mathrm{Re}_i > 400$) as:

$$N_T = 0.369\left(3\,\mathrm{Re}_f\right)^{1/3}$$

(6)

Brauer (cited in [2]) gave an empirical equation for turbulent flow as:

$$N_T = 3^{1/3}\,\mathrm{Re}_f^{\,8/15}\,400^{-1/5}$$

(7)

while Fulford [2] related the falling film thickness by dimensional analysis for channel flow, of slopes between 7.5° - 9.0°, over the range $30 < \mathrm{Re}_f < 300$, in the form:

$$N_T = 1.28 (\sin \psi)^{-0.065} \operatorname{Re}_f{}^{0.331}$$

(8)

The equations mentioned above can be applied for quasi parallel-sided film i.e. for small film thickness flow inside or outside vertical tubes. While in straight inclined tubes and spiral tubes larger film thickness is expected but they differ in effect of centrifugal force on film in spiral tubes.

Most of the recent published works on helical pipes were mainly related to fully filled fluid flow in pipes and hardly any work was found dealing with film flow.

The latest work is concerned with the pipe flow belonging to Yamamoto and co-workers [7-9]. They studied the laminar and turbulent flow through helical coils. Their numerical and experimental data concluded a negligible effect of torsion on the flow within the range of their experimental data.

Several other workers investigated the emergence of turbulence region in flow through coiled pipes, both experimentally [8-11] and numerically [12,13]. The coil curvature seems to increase the value of the Reynolds number required to attain fully turbulence flow. Difficulties were experienced in locating the transition region in the fully filled flow in helical pipes.

Vashisth et al. [14] published an intensive review on the performance of curved tubes for heat and mass transfer for fully filled flow.

Gupta et al. [15] studied the effect of coil pitch and coil diameter on the friction factor for five types of fully filled coils with different radii variation developed for Newtonian fluid. They formulated an empirical Equation (9) based on their experimental data.

$$f = \left(\frac{16}{\operatorname{Re}} \right) \left(1 + a N_{Ge}^b \right)$$

(9)

where a & b are constants and N_{Ge} is the Germano number that is equal to Re multiplied by coil curvature.

$$\text{The Coil Curvature} = \frac{\pi^2 \left(D_c / d_t \right)}{\left[\pi \left(D_c / d_t \right) \right]^2 + \left(p / d_t \right)^2}$$

They concluded that Equation (9) predicts the observed coil friction factor values to within ±10%.

Film flow in curved tubes has not been given any attention in recent literature. The present work studies the thin film flow in curved tubes and extends its findings to large film thickness. The investigation covers experimental and theoretical film flow in both straight inclined and spiral tubes and how the two systems can be related. The work also discusses the case of film flow over flat surfaces.

EXPERIMENTAL DESIGN

The apparatus used in this work is shown schematically in Figure 1, the system consists mainly of a constant head tank A, transparent flexible tube B, storage tank C and centrifugal pump D to circulate the distilled water in the system.

The testing part is made of flexible tube B having an elliptic cross section with an inside minor axis of 1.510 cm and inside major axis of 1.888 cm and a thickness of 0.412 cm.

It is known from literature that films are practically smooth for angles of inclination less than 5° over a wide range of Re_i and would have limited ripples at higher range of Re_i. For this purpose, angles of inclination between 2.5° - 5° were studied.

In straight inclined tube experiments, the flexible tube is mounted on a flat steel plate. The far end of the tube away from the entrance is supported on a horizontal stand. It can be moved up or down to a suitable distance that can be conveniently measured by a traveling microscope in order to set the required include angle (2.5°, 3°, $\cdots$ 5°).

The length of the developing region was estimated to be about 20 film thickness [16].

Hence, the film thickness is measured at 50 cm distance away from the entrance of the tube to insure that all measurements are made in the fully developed region.

A special micrometer, as shown in Figure 2, is used to measure the film thickness through holes in upper part of the flexible tube B. It consists of two micrometers arranged so that the vertical and horizontal distances of the cross sectional area of the flow can be measured. The whole arrangement is supported by a vertical stand, which allows the two micrometers to move to the required position near the system.

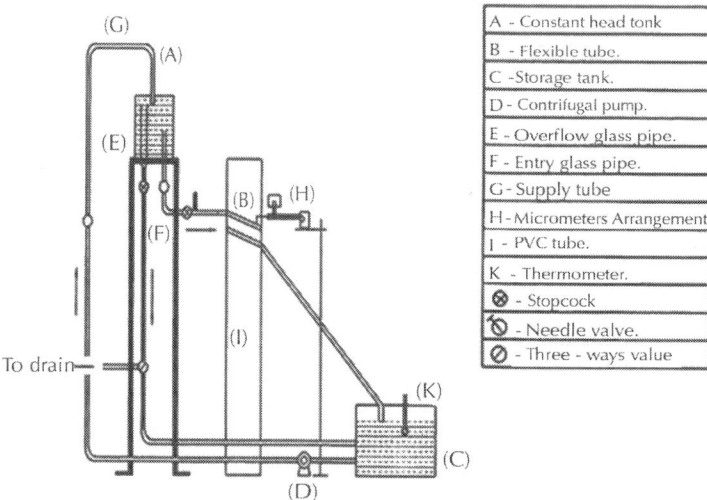

Figure 1: Chematic diagram of the apparatus used in film thickness measurement.

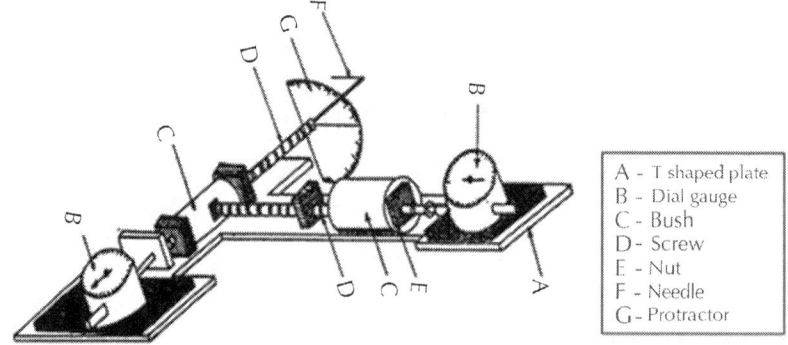

Figure 2: Micrometers arrangement.

In spiral tube experiments, the helical is made up by wrapping the flexible tube B around the PVC tube I. Three tubes I of different diameters are selected. The flexible tube B is wrapped at the same angles of inclination as mentioned above. The curvature of the tube has a calming action, the developing region in curved tubes is expected to be shorter than that in straight tubes.

THEORY AND MODEL

Film Flow in Inclined Circular Tube

The cross section of circular tube and a liquid film are shown in Figure 3. It can be obtained from trigonometric relations that:

$$b = a \frac{\cos\left(\dfrac{\alpha}{2}\right)}{\cos\left(\theta - \dfrac{\alpha}{2}\right)}$$

(10)

The Navier-Stokes equation in cylindrical coordinates for smooth, steady, laminar film flow can be reduced to the following equation:

$$\frac{d^2U_z}{dr^2} + \frac{1}{r}\frac{dU_z}{dr} = -\frac{g\sin\psi}{\upsilon}$$

(11)

This differential equation can be solved by using the following boundary conditions:

$$\left.\begin{array}{l} \text{B. C. 1}: U_z = 0 \qquad \text{at } r = a \\ \text{B. C. 2}: dU_z/dr = 0 \quad \text{at } r = b \end{array}\right] \text{ at all } \theta$$

The solution below represents the known semi-parabolic relation in the r-direction.

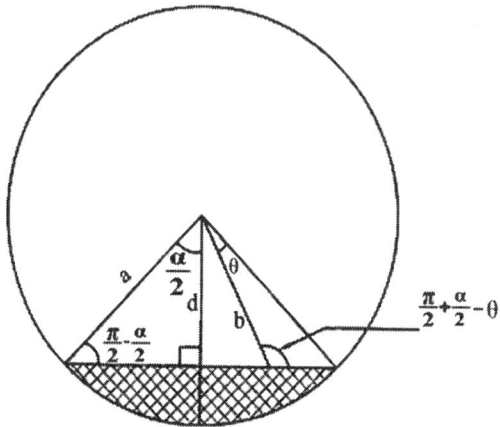

Figure 3: Liquid film in circular tube.

$$U_z = \frac{g\sin\psi a^2}{4\upsilon}\left[1-\left(\frac{r}{a}\right)^2+2\left(\frac{b}{a}\right)^2 In\left(\frac{r}{a}\right)\right].$$

(12a)

While the film surface velocity is represented by:

$$U_z\Big|_s = \frac{g\sin\psi a^2}{4\upsilon}\left[1-\left(\frac{b}{a}\right)^2+2\left(\frac{b}{a}\right)^2 In\left(\frac{b}{a}\right)\right].$$

(12b)

The average velocity is obtained from the following relationship:

$$U_{a\upsilon}=\frac{\int\limits_{0}^{\alpha}\int\limits_{b}^{a}U_z r dr d\theta}{\int\limits_{0}^{\alpha}\int\limits_{b}^{a} r dr d\theta}$$

(13)

Submitting the expression of U_z in Equation (13) and carrying out the integration and then using the trigonometric identities would result in the following equation:

$$
\begin{aligned}
U_{av} = \frac{g \sin \psi a^2}{4v} & \left[\frac{2}{a^2 (\alpha - \sin \alpha)} \left(\frac{a^2}{4} \right) \right] \\
& \cdot \left\{ \alpha - \left(\frac{16}{3} \right) \left(\frac{\alpha}{2} \right) \cos^4 \left(\frac{\alpha}{2} \right) \right. \\
& \left. - \left[2 \cos \left(\frac{\alpha}{2} \right) \sin \left(\frac{\alpha}{2} \right) \left(\frac{23}{9} - \frac{38}{9} \cos^2 \left(\frac{\alpha}{2} \right) \right) \right] \right\}
\end{aligned}
$$

$$(14)$$

Since $a^2 (\alpha - \sin \alpha)/2$ represent the cross-sectional area of the film, the volumetric flow rate can be given as:

$$
\begin{aligned}
Q = \frac{g \sin \psi a^2}{4v} \left(\frac{a^2}{4} \right) & \left\{ \alpha - \left(\frac{16}{3} \right) \left(\frac{\alpha}{2} \right) \cos^4 \left(\frac{\alpha}{2} \right) \right. \\
& \left. - \left[2 \cos \left(\frac{\alpha}{2} \right) \sin \left(\frac{\alpha}{2} \right) \left(\frac{23}{9} - \frac{38}{9} \cos^2 \left(\frac{\alpha}{2} \right) \right) \right] \right\}
\end{aligned}
$$

$$(15)$$

By comparing Equation (2), which is for the average velocity in the conventional two dimensional film flows with the derived Equation (14) which is for the average velocity in the inclined tube, it can be considered from their identity that the film thickness in inclined tube is defined by Equation (16):

$$
\begin{aligned}
\delta^2 = \frac{a^2}{2 (\alpha - \sin \alpha)} & \left\{ \alpha - \left(\frac{16}{3} \right) \left(\frac{\alpha}{2} \right) \cos^4 \left(\frac{\alpha}{2} \right) \right. \\
& \left. - \left[2 \cos \left(\frac{\alpha}{2} \right) \sin \left(\frac{\alpha}{2} \right) \left(\frac{23}{9} - \frac{38}{9} \cos^2 \left(\frac{\alpha}{2} \right) \right) \right] \right\}
\end{aligned}
$$

$$(16)$$

It is obvious that the film thickness is a function of both radius and central angle of the flow. Hence, the average velocity can be expressed as:

$$U_{av} = \frac{g \sin \psi}{4v} \delta^2$$

(17)

Further rearrangement of Equation (17) in terms of the dimensionless Nusselt film thickness and film Reynolds number would results in:

$$N_T = \left(4 \operatorname{Re}_f\right)^{1/3}$$

(18)

As would be expected the flow in inclined circular tubes (Equation (18)) predicts higher Nusselt film thickness than in case of a flow over inclined flat plate (Equation (5)); since, the geometrical shape of the former offers higher wetted perimeter ($H_R/H_D < 1$) than in the latter case ($H_R/H_D = 1$) for the same free (exposed) surface.

It is more convenient to manipulate the experimental data for the flow in elliptic cross-section in terms of hydraulic radius rather than film thickness.

Applying the least squares fit to Equation (14) and the definition of hydraulic radius (refer to Figure 4) gives.

$$U_{av} = \frac{g \sin \psi}{4v} \left(2.0925 H_R^2\right)$$

(19)

With a standard deviation of 4.578×10^{-3} and 0.94% average percentage error:

Where

$$\delta = 1.4465 H_R$$

(20)

The Film Friction Factor

As the gas phase is practically stationary in the system, negligible drag force can be assumed at the free surface of the film. Therefore, the film weight is supported by the shear stress at the wall:

$$\tau_w = \delta \rho g \sin \psi$$

(21)

By substituting the value of d from Equation (17) and using the definition of Re_f, Equation (21) will be:

$$\tau_w = \rho \left(4v^2 g^2 \sin^2 \psi \right)^{1/3} Re_f^{\;1/3}$$

(22)

By defining the friction factor in the following form:

$$\tau_w = f \rho \frac{U_{av}^2}{2}$$

(23)

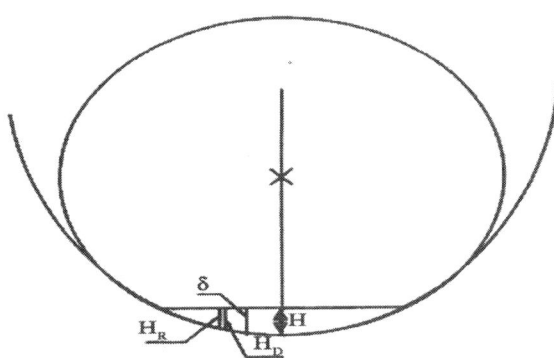

Figure 4: Hydraulic depth, hydraulic radius and film thickness at Re_f = 303, angle of inclination 3°.

and using the expression for U_{av} from Equation (17), the following equation is obtained:

$$f\left(\frac{\rho}{2}\right)\left(\frac{vg\sin\psi}{4}\right)^{2/3}\mathrm{Re}_f^{4/3} = \rho\left(4v^2g^2\sin^2\psi\right)^{1/3}\mathrm{Re}_f^{1/3}$$

(24)

Simplifying the above equation will give a correlation relating the film friction factor and film Reynolds number:

$$f = \frac{8}{\mathrm{Re}_f}$$

(25)

This is similar to the equation for laminar flow in a closed pipe. Equation (25) predicts a higher friction factor than would be predicted for liquid film falling on an inclined flat plate. Here again, the difference is due to the higher wetted perimeter for circular tubes than that of flat plates for the same free surface which would result in a higher drag force.

RESULTS AND DISCUSSION

Inclined Tube

Figure 5 shows the variation of Nusselt film thickness versus film Reynolds number. Results show that laminar and turbulent regions fall within Re_i ranging from 480 to 600. The transition region is reconfirmed at about the same Re_i in Figure 6 in which the film friction factor, calculated from Equations (21) and (23), is plotted against the film Reynolds number.

In the laminar region ($\mathrm{Re}_i < 480$) (see Figure 5), most of the experimental points lie above the theoretical line of Equation (18) with the maximum deviation falling between +30% to −7%. The least-squares fit of the experimental data gives the empirical equation:

$$N_T = 1.153\left(4\,\mathrm{Re}_f\right)^{1/3}$$

(26)

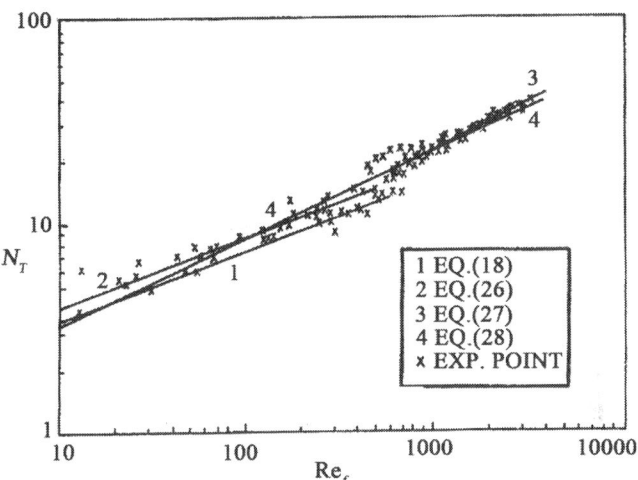

Figure 5: Inclined tube Nusselt number versus Reynolds number.

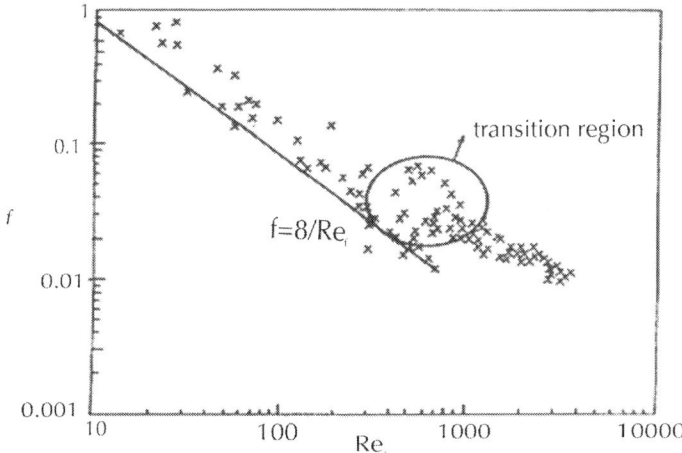

Figure 6: Film friction factor versus Reynolds number (inclined tube).

with standard deviation = 0.158 and average % error = 10.33.

While for the turbulent region (Re$_i$ > 600), the following empirical equation is obtained:

$$N_T = 0.872\,\mathrm{Re}_f^{\;8/17}$$

(27)

with standard deviation = 6.43 × 10^{-2} and average % error = 0.503.

Considering the whole experimental range of the film Reynolds number the following empirical equation is obtained:

$$N_T = 1.225\,\mathrm{Re}_f^{\;5/12}$$

(28)

with standard deviation = 0.164 and average % error = 9.64.

The three above empirical equations, are shown in Figure 7 together with Nusselt, Fiend and Brauer equations for film flow over flat surfaces. For film flow in inclined circular tubes a higher Nusselt film thickness for both the laminar and turbulent regions is predicted.

Ripple inception is observed to occur at Re$_f$ ranging from 260 to 300 within all range of the angles of inclinations used in the experimental runs. Thomas, et al. [17] observed a ripple inception at lower range of Re$_f$ i.e.between 80 - 130 for the central region of the film in a wide channel. This difference in behavior may be due to the relatively narrow tube used in this study, where surface tension would have greater influence on the flow.

Helical Tube

A flexible transparent plastic tube is wrapped around a plastic cylinder at a predetermined angle of inclination, as in Figure 1. During the experimental runs, it was observed that the instability in film behavior increases with film flow rate and helix curvature.

As the flow rate is reduced, small cross waves were observed moving from the inner to the outer walls of the tube. These cross-waves disappear at lower flow rates and are replaced by small random ripples.

At this stage four to five readings were recorded for film thickness.

First and second order polynomials were used to represent surface profile of the film and to extrapolate the surface shape near the inner and outer walls of the tube.

Most of the results suggested that first order polynomial correlation applied especially at low surface gradient. A typical representation of experimental runs is shown in Figure 8, which shows the unaxisymmetrical behavior of the film in helical tubes.

For each run the hydraulic radius, hydraulic depth and cross-sectional area of the flow are calculated. When the surface of the film was inclined, the surface length and the cross-sectional area of the flow were calculated using the analytic integration method for the surface equation.

In Figures 9 and 10, Nusselt film thickness is plotted against film Reynolds number. The transition region is not so distinguished as an inclined tube. In the helical tube gradual change of flow behavior confuses the limits of the transition region. This behavior is similar to itscounter part of a full pipe flow.

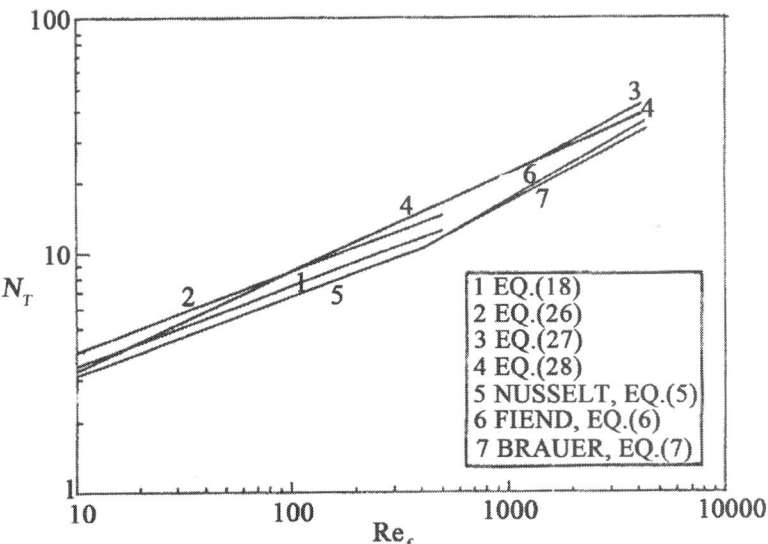

Figure 7: Comparison of present relation of film flow in inclined tube with that of inclined flat plate.

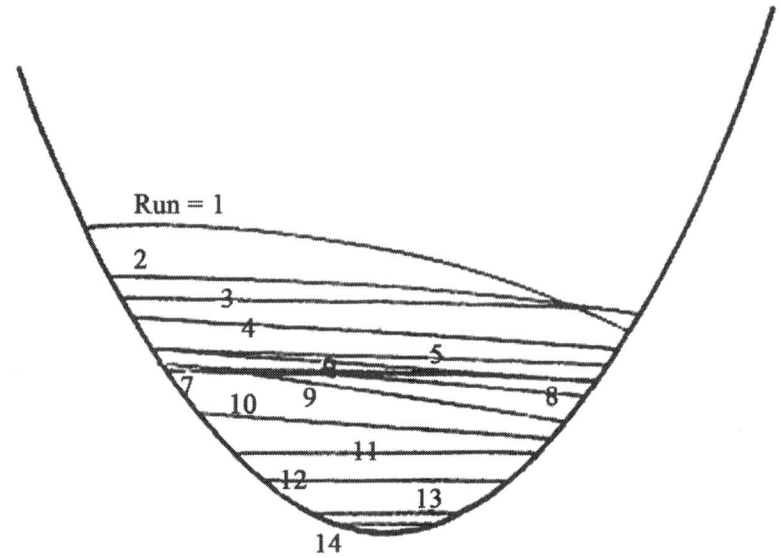

Figure 8: Film surface profile in elliptic cross sectional tube (polynomial approximation).

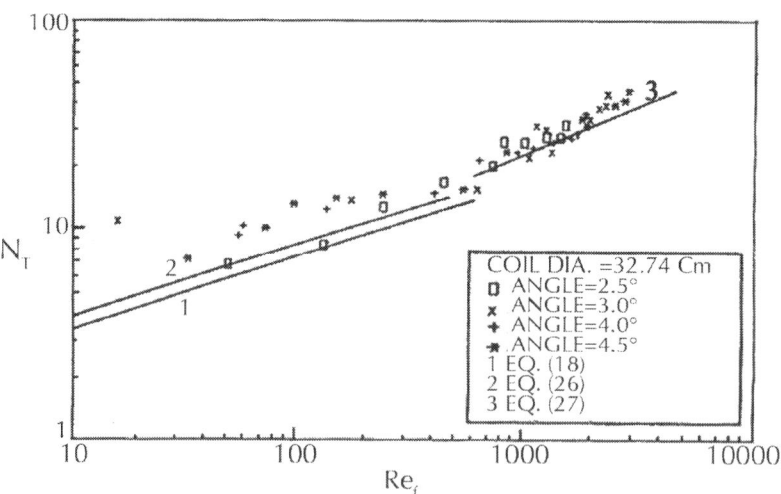

Figure 9: Helical tube Nusselt number versus Reynolds number (pitch effect at constant curvature).

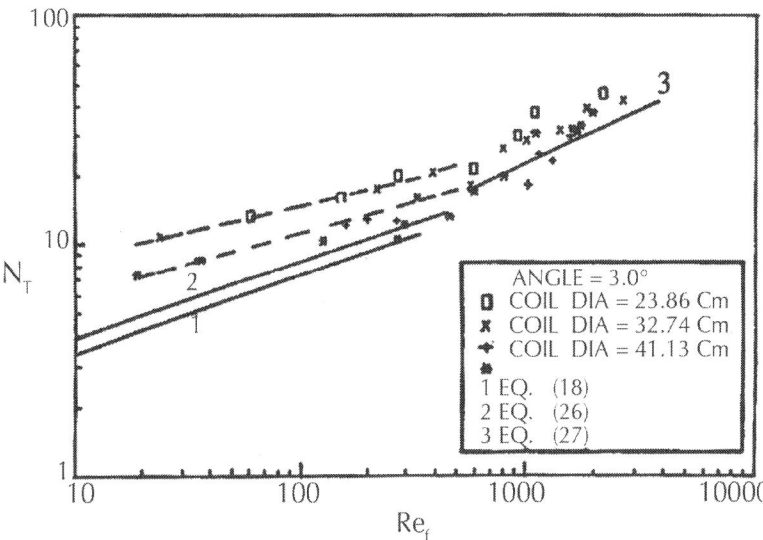

Figure 10: Helical tube Nusselt number versus Reynolds number (Curvature effect at constant pitch of 300 angle).

In the laminar region, the effect of pitch (angle of inclination) is clearly evident (see Figure 9) at certain film Reynolds number.

For turbulent flow, the action of coil diameter (curvature) and the action of centrifugal force would mostly reduce by inertia force and by the development of eddies. The effect of higher curvature can be observed while the pitch effect is overlapped by other effects.

In Figures 11 and 12, the Nusselt film thickness is plotted against the Dean number (De_i) which combines the effects of both curvature and film Reynolds number. The figures reveal that the spread in the experimental results is reduced as compared to Figures 9 and 10.

In order to relate Nusselt film thickness for coiled tubes to that of straight tubes, Figures 13 and 14 are plotted in terms of $N_{T(c.tube)}/N_{T(st.tube)}$ versus film Dean number. Equation (28) is applied to calculate $N_{T(st.tube)}$. It can be observed that most of the experimental points lie between value of $N_{T(c.tube)}/N_{T(st.tube)}$ of 1 - 2. This ratio tends to decrease gradually with the increase of De_i at certain coil diameter.

By the least squares fit, the following empirical correlation is obtained:

$$\frac{N_{T(c.tube)}}{N_{T(st.tube)}} = 11.0 De_f^{-0.1}\left(\frac{a}{R}\right)^{0.55}$$

(29)

with Standard Deviation = 0.106 and average % error = 9.56.

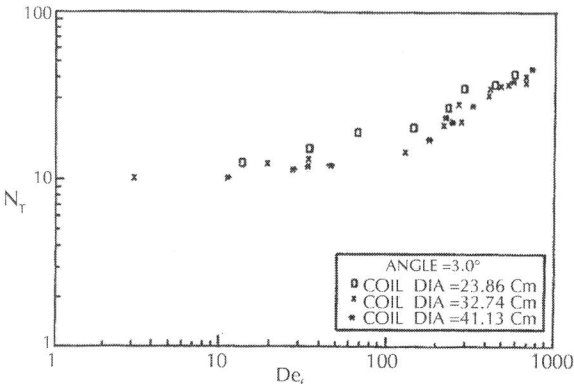

Figure 11: Helical tube Nusselt number versus Dean number(curvature effect at constant pitch of 3.00 angle).

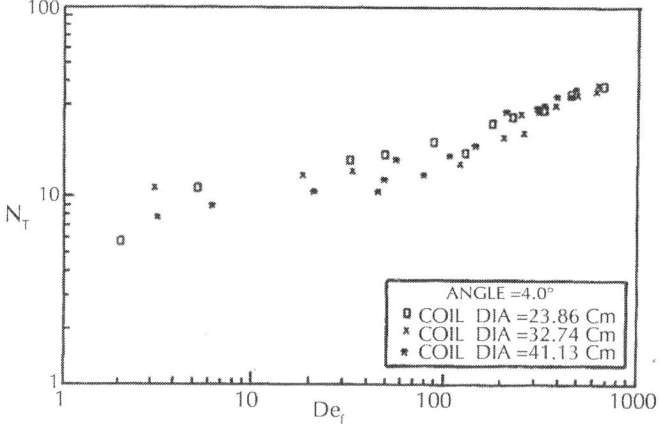

Figure 12: Helical tube Nusselt number versus Dean number(curvature effect at constant pitch of 4.00 angle).

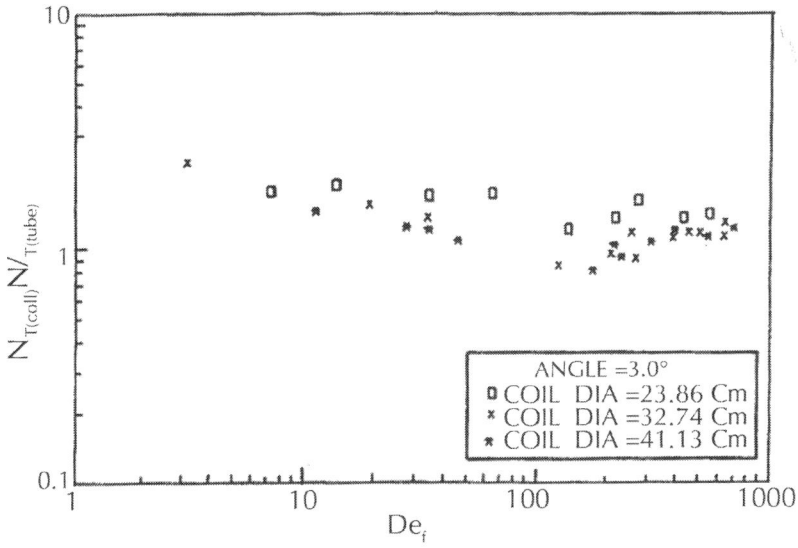

Figure 13: Ratio of Nusselt number of helical tube over Nusselt number (curvature effect at constant pitch of 3.00 angle).

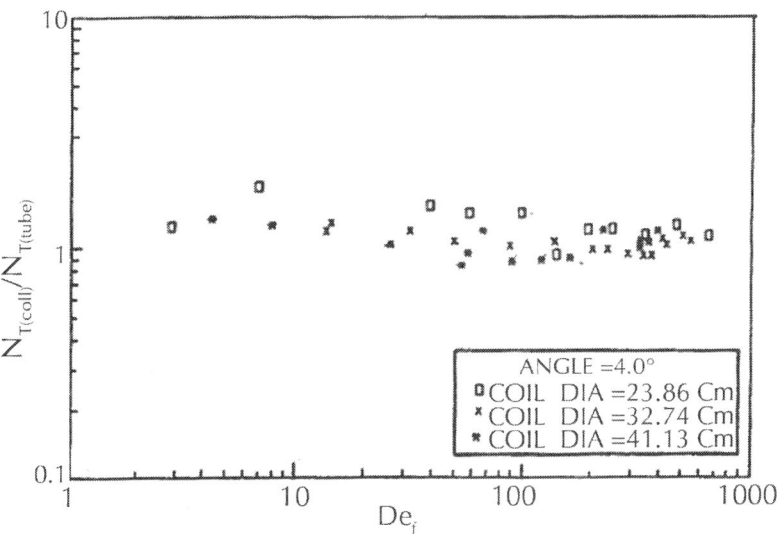

Figure 14: Ratio of Nusselt number of helical tube over Nusselt number (curvature effect at constant pitch of 4.00).

CONCLUSIONS

- Systems that have their film thickness vary peripherally can be characterized by some average film thickness. This is a function of hydraulic radius defined by the film boundary.

- In helical tubes, the film thickness increases with the increased curvature accompanied by a reduction in average velocity. Within the range of the obtained data, coiling effect may account for the maximum increase in film thickness of 70% over that of the inclined tube. But, this increase diminishes gradually in the turbulent region, where the secondary flow intensity is small relative to axial flow intensity.

- Pitch within two to three tube diameters in length has no appreciable effect on film thickness at low tube curvature.

- The transition region in a straight inclined tube is marked between $Re_i = 480 - 600$, which is higher than the reported range for vertical tubes ($Re_i = 350 - 500$). In helical tubes, the transition region is not so distinct (within experimental data) as it is in inclined tube 5) Results are correlated empirically for the cases of straight inclined and helical tubes. The former is expressed in terms of N_T as function of Re_i in the laminar and turbulent regions. While for the latter the empirical correlation is represented in terms of the ratio of N_T to coil over N_T for straight inclined tubes as function of De_f.

REFERENCES

1.   V. T. Chov, "Open-Channel Hydraulic," McGraw-Hill, Kogakusha, 1959.

2.   G. D. Fulford and T. B. Drew, "Advances in Chemical Engineering," Academic Press, New York, Vol. 5, 1964.

3.   R. B. Bird, W. E. Steward and E. N. Lightfoot, "Transport Phenomena," John Wiley, New York, 1960.

4.   H. Jefferys, "The Flow of Water in an Inclined Channel of Rectangular Section," Philosophical Magazine, Vol. 49, 1925.

5.   C. M. Cooper, T. B. Drew and W. H. McAdams, Transactions of the American Institute of Chemical Engineers, Vol. 30, 1934.

6.   C. G. Kirkbride, Industrial & Engineering Chemistry Research, Vol. 26, 1934.

7.   K. Yamamoto, S. Yanase and T. Yoshida, "Torsion Effect on the Flow in a Helical Pipe," Fluid Dynamics Research, Vol. 14, 1994, pp. 259-273. doi:10.1016/0169-5983(94)90035-3.

8.   K. Yamamoto, T. Akita, H. Ikeuchi and Y. Kita, "Experimental Study of the Flow in a Helical Circular Tube," Fluid Dynamics Research, Vol. 16, No. 4, 1995, pp. 237-249.doi:10.1016/0169-5983(95)00022-6.

9.   K. Yamamoto, A. Aribowo, Y. Hayamizu, T. Hirose, K. Kawahara, "Visualization of the Flow in a Helical Pipe," Fluid Dynamics Research, Vol. 30, No. 4, 2002, pp. 251-267.doi:10.1016/S0169-5983(02)00043-6.

10.   D. R. Webster and J. A. C. Humphrey, "Experimental Observation of Flow Instability in a Helical Coil," Fluid Engineering, Vol. 115, 1993, pp. 436-443. doi:10.1115/1.2910157.

11.   A. Cioncolini and L. Santini, "An Experimental Investigation Regarding the Laminar to Turbulent Flow Tansition in Helically Coiled Pipes," Experimental Thermal and Fluid Science, Vol. 30, No. 7, 2006, pp. 367-380. doi:10.1016/j.expthermflusci.2005.08.005.

12.   T. J. Hüttl and R. Friedrich, "Influence of Curvature and Torsion on Turbulent flow in helically Coiled Pipes," International Journal of Heat and Fluid Flow, Vol. 21, No. 3, 2000, pp. 345-353.

13.   T. J. Hüttl and R. Friedrich, "Direct Numerical Simulation of Turbulent Flows in Curved and helically Coiled Pipes," Computers & Fluids, Vol. 30, No. 5, 2001, pp. 591-605.doi:10.1016/S0045-7930(01)00008-1.

14.   V. Vashisth, V. Kumar and K. D. P. Nigam, "A Review on That Potential Applications of Curved Geometries in Process Industry," Industrial & Engineering Chemistry Research, Vol. 47, No. 10, 2008, pp. 3291-3337. doi:10.1021/ie701760h.

15.   R. Gupta, R. K. Wanchoo and T. R. M. J. Ali, "Laminar Flow in Helical Coils: A Parametric Study," Industrial & Engineering Chemistry Research, Vol. 50, No. 2, 2011, pp. 1150-1157. doi:10.1021/ie101752z.

16.   B. Atkinson and R. L. McKee, Chemical Engineering Science, Vol. 19, 1964.

17.   W. J. Thomas, M. S. Ray and E. W. Palmer, "Physical Absorption of Carbon Dioxide in Water Flowing in an Inclined Cell," Chemical Engineering Communications, Vol. 2, No. 3, 1976, pp. 121-134. doi:10.1080/00986447608960454.

Computational Tools to Support Ethanol Pipeline Network Design Decisions

Gustavo Dias da Silva*, Virgílio José Martins
Ferreira Filho, Laura Bahiense

Production Engineering Program, Federal University of Rio de Janeiro, Rio de Janeiro, Brazil

ABSTRACT

This paper considers the pipeline network design problem (PND) in ethanol transportation, with a view to providing robust and efficient computational tools to assist decision makers in evaluating the technical and economic feasibility of ethanol pipeline network designs. Such tools must be able to address major design decisions and technical characteristics, and estimate network construction and operation costs to any time horizon. The specific context in which the study was conducted was the ethanol industry in São Paulo. Five instances were constructed using pseudo-real data to test the methodologies developed.

INTRODUCTION

People use networks daily (sometimes unwittingly) in a wide range of contexts. Telephony networks, logistics networks for freight transportation and distribution networks (water, energy) provide services that are essential to modern life and, therefore, must be properly designed to ensure service quality and meet user demands.

Brazil is projected to be both a major producer and supplier of biofuels, particularly biodiesel and ethanol. These fuels are gaining importance on the world stage in view of the environmental, technological and economic impacts caused by global dependence on fossil fuels, particularly oil [1].

However, Brazil's transport logistics system still rests largely on its highway network, which—although economically competitive—is not safe for transporting fuel, especially in large volumes. A paradigm shift must take place and Operations Research can help in that process.

This study aims to provide robust and efficient computational tools to assist decision makers in evaluating the technical and economic feasibility of ethanol pipeline network designs. Such tools must be able to address major design decisions and technical characteristics, and estimate network construction and operation costs to any time horizon.

The paper is organized as follows: Section 2 presents the context that motivated the research; Section 3 describes the problem under study, characterizing it in technical and theoretical terms; Section 4 presents the optimization approaches taken to solving the problem; Section 5 reports the results and conclusions; and Section 6 offers some final remarks on the study.

CONTEXT

Current economic growth rates have generated increasing demand for oil. Consequently, market laws exert strong pressure on oil prices, which fluctuate, significantly increasing the cost of living. In parallel with this, the environmental effects of global warming are an issue globally. Pollutant emissions from burning fossil fuels aggravate the phenomenon, drawing criticism and spurring the search for viable alternatives [1]. Technological development has also contributed significantly to this substitution. "Flex-fuel" vehicles have reduced fossil-fuel consumption by offering consumers freedom of choice [2].

Ethanol will play an important role in this process: firstly, because large-scale production costs have been competitive with those of oil since 2004, and secondly, because it is a cleaner energy source [3]. Brazil has become critical in this process, because it has the experience and natural conditions for producing ethanol from sugarcane, and has invested strongly and achieved technological improvements since the start of the Proálcool program [4,5]. Figure 1 shows Brazil's output by region in crop year 2008/2009 [6].

Brazil is now the world's second-largest producer [4]. According to the Sugarcane Industry Union, [6], production reached 27.5 million cubic meters in crop year 2008/ 2009, and is increasing each season. Figure 2 shows the trend [6]. However, studies have found that bottlenecks in Brazil's infrastructure significantly affect fuel logistics, burdening links in the supply chain and raising endproduct prices [7].

This result is worrying given that production is not evenly distributed across Brazil. Though concentrated mostly in the southeast [8], due mainly to different production technologies [3], a logistics network is essential to collecting it at lowest possible cost. Pipelines are a particularly appropriate transport mode for such transfers, as they always attain large volumes (Figure 3). Motivated by this context, this study examines the logistics of fuel. The particular problem addressed is how to design a pipeline network to collect the ethanol produced by a set of regions and route it to a pre-determined destination region. Pipelines were chosen for their high reliability and economic competitiveness [9]. São Paulo State was chosen since it is currently Brazil's largest ethanol producer [6], thus justifying investments in the sector. Figure 2 illustrates its ouptut.

Designing such networks is a complex task due to the large number of technical and economic requirements to be considered simultaneously. Proper project management is imperative, and projects are commonly divided into conceptual and hierarchically dependent phases that characterize their life cycle [10]. A pipeline project consists of the following stages [9]:

- Planning: early stage of the potential enterprise, where technical, economic and social data are collected to assist feasibility evaluation.

- Conceptual Engineering: identifies project technical and economic feasibility, and sets the basic and detail engineering agenda.
- Basic Engineering: the basic design is developed and detailed to consolidate engineering aspects before any procurement and implementation expenditures are made.
- Implementation: step execution and control. Activities typical of this stage include construction and assembly, commissioning and conditioning.
- Operation and Maintenance; Decommissioning: operation of pipeline is terminated because it is obso lescent, supplanted by other systems or no longer of interest to the owner.

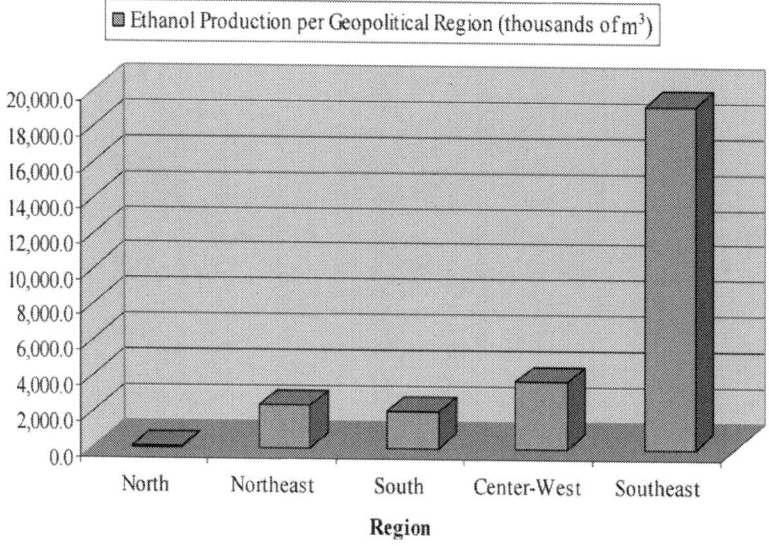

Figure 1: Ethanol production by geopolitical region, 2008/2009 harvest.

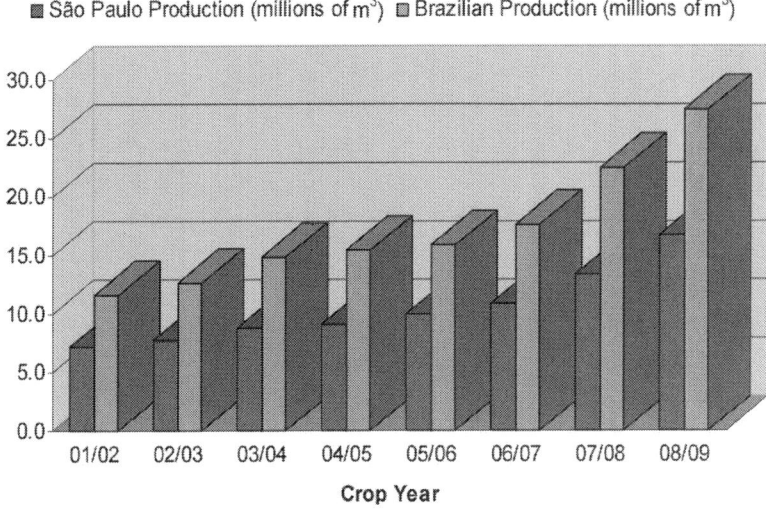

Figure 2: Ethanol production in São Paulo and Brazil.

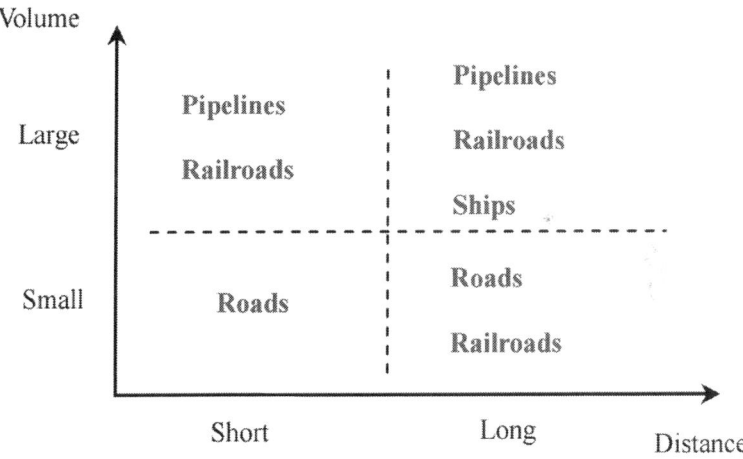

Figure 3: Transport mode comparison matrix.

From an Operations Research viewpoint, meanwhile, Minoux [11] points out that, given a set of nodes, designing a network consists of constructing and/or installing better connections between some pairs of these nodes in compliance with certain operational specifications that govern the application's behavior. That definition covers the planning and conceptual engineering phases mentioned above, where the main concern is to assess the strategy's competitiveness, and where the level of technical detail is still low. This study intends to develop computational tools to assist in that endeavor. Silva [12] offers a brief literature review on the PND (and correlated problems). Distribution network designs are frequent, and entail deciding pipeline diameters, pump capacities, reservoir elevation profiles and duct flows, satisfying constraints, ensuring hydraulic energy balance and meeting demand. Such decisions and constraints are handled in an integrated or hierarchical manner, frequently taking topology as given. None address the problem examined here. The most similar, offered by Brimberg et al. [13], consists in designing an optimal pipeline network topology to transport output from oil fields in Gabon by connecting oil wells to a port.

Investment and operating costs tend to be of different orders of magnitude [9]. In the context considered here, however, all information about total project cost is important, since roads historically form the backbone of Brazil's logistics system [2] and are always an economically competitive alternative. In addition, depending on the time horizon considered, operating costs may influence design decisions. The combinatorial nature of the problem, coupled with the nonlinear equations of fluid dynamics, requires the use of a mixed-integer nonlinear program (MINLP) approach, which is solved via exact and heuristic algorithms.

PIPELINE NETWORK DESIGN

This section will give an overview of ethnol pipeline network design, indicating the key theoretical and technical characteristics. The problem is formally stated, presenting the technical decisions that are being addressed and the main assumptions. Next a cost analysis is made of pipeline network projects. Finally, the problem is formulated mathematically.

Fluid Dynamics Considerations

Flows

Flows are caused by the action of a shear stress on a fluid initially at rest. Temporal analysis of a flow allows it to be classified into one of two regimes, namely the transitional (early stage) regime, in which flow velocity, density and temperature parameters at any point vary with time, and the steady-state (perennial stage) regime in which these parameters do not vary with time.

Steady-state flows, in turn, can be classified into two types: laminar, in which fluid particles move along parallel streamlines (paths) without macroscopic interaction between the layers; and turbulent, in which fluid particles move along irregular trajectories with major macroscopic interaction. The dimensionless Reynolds number is used to indicate whether the flow is laminar or turbulent [14].

In ducts, flow may occur under the predominant, but not exclusive, action of two different mechanisms: gravity or a pressure gradient. The main characteristic of flow under gravity, known as open-channel flow, lies in the fact that the fluid flows without completely filling the duct. In the case of flows promoted by a pressure gradient. The ducts are completely filled by the fluid transfer.

Finally, it is possible to characterize two regions in flow along a pipeline: the entrance region, where the velocity field profile is not yet fully developed, and the fully-developed flow region, where the velocity field is the same at any cross-section. The fully-developed region may not exist, but generally, for a pipeline length L and diameter ϕ, if ,$L \gg 30\phi$ it can be assumed that the steady state has been reached.

Energy Issues

Pipeline networks comprise ducts and other components such as valves, bends, fittings, pumps, turbines and tanks. All these elements contribute to system energy dissipation. There are two types of losses, one due to friction between duct walls and the outer layers of the fluid, and the other due to geometric changes in cross-section profiles along the pipeline route. The D'Arcy-Weisbach Equation enables losses to be calculated for incompressible, permanent and fully-developed flows:

$$E_f = \frac{8f^\phi}{g\pi^2} \frac{LQ^2}{\phi^5},$$

(1)

where E_f is the pressure drop along the pipeline whose transferred flow is Q, π the parameter pi, g the gravity (m/s^2) and f^ϕ the friction factor.

Flow machines are mechanical devices capable of extracting energy from fluids (turbines) or adding energy to them (pumps) by dynamic interactions between the device and the fluid [14]. Pumps, in turn, are described mainly by characteristic curves that relate the load provided and the operation flow. Figure 4 exemplifies this. Another important parameter in describing pump energy behavior is the Net Positive Suction Head (NPSH), which relates to cavitation (vapor bubbles inside the device). Commonly, the pressure in the pump's power section (input) is low and, when this value reaches the vapor pressure of the fluid, a fraction of the liquid evaporates. This phenomenon leads to loss of efficiency and structural damage.

In 1738 Bernoulli formulated the basic equation governing mechanical energy conservation in a fluid flow system. Generalized a few years later, the energy balance for an isothermal, confined, steady and turbulent flow of a viscous fluid that receives work from, and performs work on, flow machines is given by:

$$\frac{\Delta P}{\gamma} + \frac{\Delta V^2}{2g} + \Delta Z - E_S - W_S + W_{MF} = 0,$$

(2)

where γ is the density of the transferred fluid. It represents the sum of pressure (1st term), kinetic (2nd term) and gravitational potential (3rd term) energies in the system, the total power loss (E_s), the work done by the system on its border (W_s), and the work performed by device B on the system(W_{mf}).

Equipment Selection

Pumps to ensure flows must be selected appropriately not to overestimate or underestimate the equipment's capacity, thus avoiding unnecessary expense. The system curve is a curve relating energy flow to the load supplied to the fluid, considering the network's technical and operational features. Consider the system in Figure 5. Substituting Equation (1) into (2), between A and C:

$$W_{MF} = K_\Delta + K_S Q^2,$$

(3)

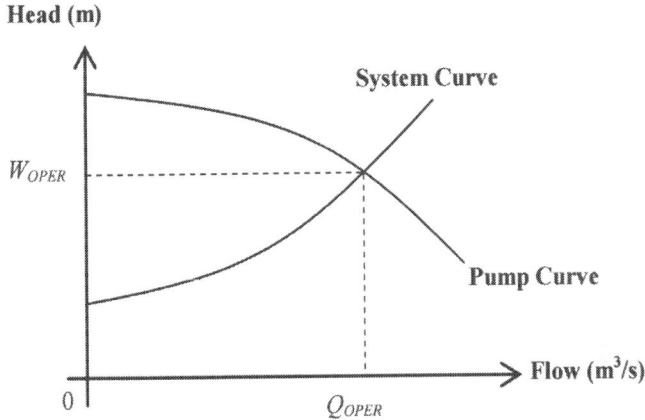

Figure 4: Equipment selection using system and pump curves.

where K_Δ and K_S are appropriate parameters. The Equation (3) characterizes the system curve, relating the energy supplied and the flow. This is shown in Figure 4 for a generic set of technical parameters.

System and pump curves are used to choose equipment. Given a system curve and an operating flow, Q_{OPER} the best-suited pump B to provide the required W_{OPER} energy is a pump whose characteristic curve intersects the system curve at the desired flow operating point (Figure 4). However, the efficiency curve of the pump should also be considered. Ideally, the operating point should be as close as possible to the highest efficiency point [14]. This latter curve will be disregarded for simplicity.

General Characteristics of the PND

Ethanol production is a seasonal activity, and the productive capacity of any region is subject to the conditions necessary for crop development.

Such issues make production minimally random in terms of constancy and amount to a given time horizon. Also sugar is produced from sugar cane, and high-demand situations impact ethanol production [15]. Proper treatment of these issues calls for stochastic approaches. Here, however, ethanol output is assumed to be known and not to vary to a given time horizon.

There is no minimum pressure requirement at the end of each section, provided the flow reaches its destination. Hence, pressures will be considered zero at these points. Economic gains also motivate this simplification. Still, assuming that the flows are promoted pressure gradients alone (Section 3.1.1), this hypothesis requires that at least one pumping station be installed at the entrance to each section. This requirement is naturally well conceived, since geographical dimensions require the use of storage tanks, which themselves result in pressure discontinueties.

In an ideal operating regime, storage tanks would be unnecessary in the producer regions, because flow balance would be automatically guaranteed. In practice, in addition to production inconstancy over time, maintenance system halts are usually necessary, requiring a minimum installed storage capacity in each region. Estimates of such capacities should consider both the randomness of the production and the system downtime schedule, leading again to a stochastic approach. Ideal balance is assumed and also that no storage capacity is required.

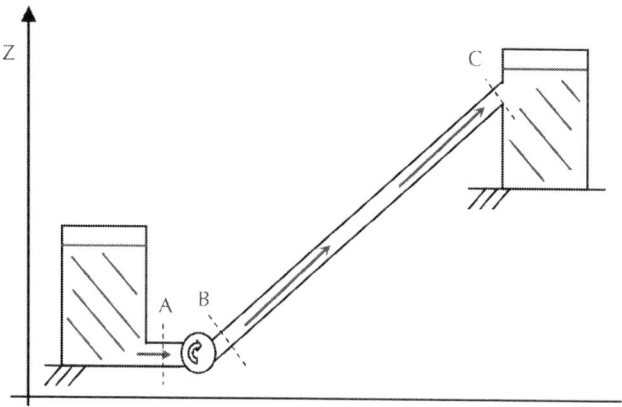

Figure 5: Typical schema of flow within a pipeline.

The essential nature of services provided by networks requires the system to be able to satisfy design demands even under failure. A redundant network ensures there are at least two distinct paths between any two network nodes. However, this solution requires construction of more pipelines than minimally required to connect all nodes in the network, which in geographically extensive networks, can represent overinvestment. The common solution in such cases is to implement a preventive maintenance policy. It is assumed therefore that the network is designed with no redundancy.

On the other hand, for a given set of technical specifications (diameter, thickness and material output), ducts are certified to withstand a certain maximum operating pressure. That relationship is given by the following equation:

$$P_{MAX} = \frac{2F\sigma e}{\phi},$$

(4)

where F is a safety factor [16], σ the circumferential tension in the pipeline, ϕ the diameter and e the thickness. The operating pressure is a function of the amount of energy supplied to the system through pumping stations. To prevent limits being exceeded, several stations are placed along sections of the network. For simplicity, however, it is assumed that only one station will be installed in each region. As a result, the maximum operating pressure will occur exactly at the entrance to each section. In a duct with constant flow and speed, zero final pressure and no flow machines, it follows from (1), (2) and (4) that:

$$Q^{\phi}_{OPER} = \frac{\pi\phi^2}{2}\left[\left(\frac{g}{f^{\phi}L}\right)\left(\frac{F\sigma e}{\gamma} + \frac{\phi\Delta Z}{2}\right)\right]^{1/2},$$

(5)

Where Q^{ϕ}_{OPER} estimates the maximum value of the flow in the duct.

Route engineering entails certain difficulties. First, it is impossible to build pipelines in straight lines, since there may be impassable terrain between points of interest. Additionally, many slopes may be present in the altitude profile between two regions and this has considerable impact on pump dimensioning. Such issues are to be addressed at data level: the former, when considering distances among the regions, and the latter, using an "equivalent" altitude difference to reflect effects of slopes along routes.

Pump dimensioning is an important issue. This is done to avoid cavitation (Section 3.1.2) by estimating these pumps' NPSH, which is a function of the height of an equivalent liquid column upstream of the pump. Estimating NPSH, however, would entail estimating the maximum capacity of the tanks upstream of the pump, which would make the mathematical model much more complex. Thus, dimensioning will be done by estimating the downstream energy.

Geographically distributed pipeline networks are subject to temperature variations. This means that fluid property values may change significantly, which may invalidate use of the equations presented in Section 3.1. Thus, flow is assumed to be isothermal. Moreover, it is assumed to occur in a turbulent regime, but in fully developed state, since the distance the flow travels in each network segment is much greater than the pipeline diameter, a necessary condition for such a hypothesis to be true (Section 3.1.1).

Finally, the flow along the network is not energyconserving (Section 3.1.2). Energy losses are usually offset using pumps in order to ensure system energy balance. Equation (2), used to guarantee this balance in the mathematical model, furnishes an estimate of the energy to be provided to the system at pumps downstream. Losses due to geometric changes will be considered using an estimator for this loss, since there is no way a priori to define the set of components installed. Finally, it is assumed that the flow does not perform work on its surroundings.

The PND Studied

The problem of designing an ethanol pipeline network, given a set of production regions, $I=\{1, 2, \ , N\}$ a set of associated flows $Q=\{Q_1, Q_2, \ , Q_N\}$ and available diameters $D = \{\phi 1, \phi 2, \cdots, \phi_{|D|}\}$, consists of selecting a subset of links between pairs of these regions to constitute the topology, an associated diameter for each of these connections and, finally, calculating the energy to be introduced into the system to ensure the flow to destination region θ, while minimizing total project cost.

Cost Analysis

Economically, there are many significant costs in a pipeline project [9]. However it is virtually impossible to deal with them simultaneously, considering the technical complexity involved. One has to select those that represent a significant proportion of project cost, those that can be adequately estimated from the decisions being taken, and those directly influenced by such decisions.

Topology decisions impact on procurement and assembly costs. The same is true of diameters, since different diameters have different prices and are more or less costly to assemble. Pumps are dimensioned (Section 3.1.3) indirectly by calculating from path, length, diameter and flow. Thus, fixed and variable pumping costs are directly influenced by the decisions taken and will be considered in the model.

The first three costs are associated with the network construction phase and the last one with the operation phase (to a time horizont). This can be expressed mathematically as:

$$CT_{RD} = CA_D + CM_D + CBF + CBV(t),$$

(6)

where CT_{RD} is total network cost; CA_D total pipeline procurement cost; CM_D total pipeline installation cost: CBF the total fixed pumping cost; and CBV(t) the variable pumping cost to operating time horizon t. CBF accounts for approximately 17% of total cost [9].

Pipe Procurement

Acquisition cost (CA_D) can be estimated using the cost per unit mass ($/kg) of the material from which they are manufactured. For this, consider the cross-section of sector tr of length L_{tr}. Defining $c\alpha^\phi$ as the cost of procuring a unit length ($/m) of pipe with internal diameter $\phi_{int} = \phi$, one obtains:

$$ca^\phi = c_M \rho_M \pi \cdot e \cdot (\phi + e),$$

(7)

That is, the unit procurement cost of length of a pipe with diameter ϕ is a function of the cost per unit mass (C_M) of the material M, the density (ρ_M), the thickness of the pipe (e) and the diameter ϕ. Thus:

$$CA_D = \sum_{tr \in \Gamma} ca^\phi L_{tr},$$

(8)

where Γ is the set of sectors that constitute the topology. The Equation (8) represents a linear approximation to the total procurement cost.

Network Construction

It is unlikely that an analytical expression exists or can be derived (as in Section 3.4.1) to estimate pipeline installation costs, as installation is a practical task that depends on many factors. However, it may be possible, by empirical analysis of historical data (past projects), to determine approximate, probably non-linear, equations for such costs. However, one can assume that the cost of installing a pipeline sector is directly proportional to its length (the longer the sector length, the greater the amount of resources needed to install it) and depends on the chosen diameter of pipe (handling pipes with different diameters can be more or less costly).

Thus, the cost of installing Γ is given by:

$$CM_D = \sum_{tr \in \Gamma} ci^\phi L_{tr},$$

(9)

where ci^ϕ is the cost of installing a unit length of pipewith diameter ϕ. More complex and realistic cost functions can be used to replace (9), although this introduces nonlinearities into the total project cost function, which would complicate the solution to the problem.

Fixed Pumping

Figure 6 represents the system curve for a sector tr of pipeline with diameter ϕ. For each of the intervals constructed by partitioning the feasible flow space, defined by Q_{MIN} and Q_{MAX}, a pump B with procurement cost C_B is selected (Section 3.1.3). This yields the step cost function represented in Figure 7, for the pressure domain defined by P_{MIN} and P_{MAX}.

The function obtained takes discrete values that depend on the pressure range sector (tr) is operating in. But selecting a reference pressure for each pressure interval, one can apply a linear regression to obtain the linear pumping cost function shown in Figure 7, defined mathematically by:

$$CBF_{tr}\left(P_{tr}^{\phi}\right) = cbf_{tr}^{\phi}\,P_{tr}^{\phi} + cbfc_{tr}^{\phi},$$

(10)

where cbf_{tr}^{ϕ} is the cost of providing a unit of pressure, $cbfc_{tr}^{\phi}$ is the fixed cost associated with operating sector tr, and P_{tr}^{ϕ} its operating pressure. The regression must satisfy the condition $CBF(P_{MIN}) > 0$ (negative costs. for viable pressure values are meaningless). Considering all the sectors that constitute τ:

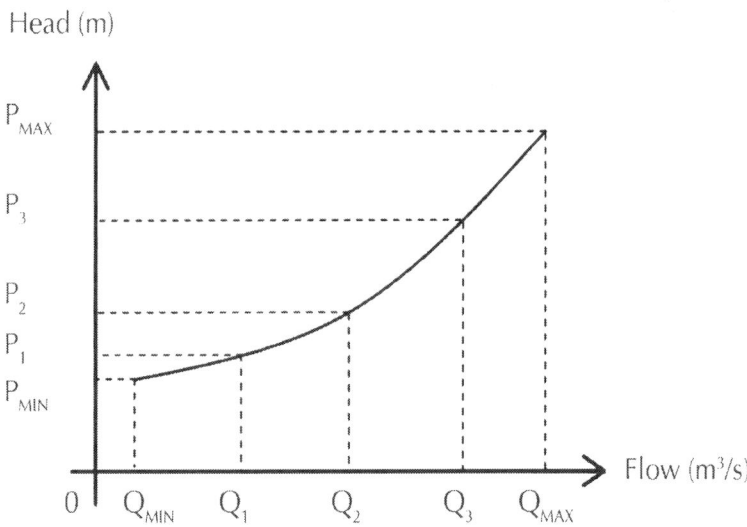

Figure 6: System curve for pipe sector tr with diameter ϕ.

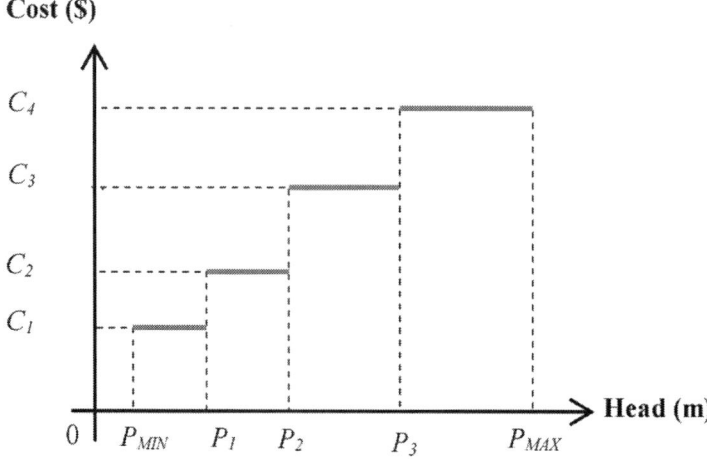

Figure 7: Fixed pumping cost step function (red) and affine function (azul) for sector tr.

$$CBF = \sum_{tr \in \Gamma} \left(cbf_{tr}^{\phi} P_{tr}^{\phi} + cbfc_{tr}^{\phi} \right)$$

(11)

Regressions for higher degree polynomials can be performed, resulting, however, in the inclusion of nonlinearities in the project cost function.

Variable Pumping

The procedure for estimating the variable pumping cost is the same, except that the costs C_B associated with the selected pumps represent the cost of operating them to time horizon t. Thus, the operating cost of τ for t can be estimated by:

$$CBV(t) = \sum_{tr \in \Gamma} \left(cbv_{tr}^{\phi} P_{tr}^{\phi} + cbvc_{tr}^{\phi} \right),$$

(12)

where cbv_{tr}^{ϕ} is the cost of providing a unit of pressure in sector tr, $cbvc_{tr}^{\phi}$ is the fixed cost of operating the sector, and P_{tr}^{ϕ} the operating pressure. The resulting expression is not time-dependent, but is addressed indirectly in (12). The positivity issue also applies here, as do higher-order regressions.

Total Cost

Substituting (8), (9), (11) and (12) in (6), the overall cost CT_τ of network τ may be estimated by equation:

$$CT_\Gamma = \sum_{tr \in \Gamma} \left(ca^\phi L_{tr} \right) + \sum_{tr \in \Gamma} \left(cbf_{tr}^\phi P_{tr}^\phi + cbfc_{tr}^\phi \right)$$
$$+ \sum_{tr \in \Gamma} \left(ci^\phi L_{tr} \right) + \sum_{tr \in \Gamma} \left(cbv_{tr}^\phi P_{tr}^\phi + cbvc_{tr}^\phi \right) \qquad (13)$$

Note that (13) was derived considering only one diameter value ϕ, but will be generalized later.

Mathematical Modeling

The set of producing regions and candidate links can be represented by a complete digraph, where vertices are regions and edges the links between them. Values for distances and inter-region height differences are associated with edges. Similarly, a graph can be associated with the designed network structure. The graph that satisfies the coverage condition for all regions with the fewest connections is a spanning tree (ST) [17], which is adopted as a solution. The formulation by Gavish [18] will be taken as the basis for modeling.

Sets, Indexes and Nomenclature

The set of regions is represented by I and diameters by D. The reduced set of regions is defined by $\dot{I} = I / \{\theta\}$. The indexes i, j and k and index d represent the regions and the diameters, respectively. An edge or link (I, j, d) interconnecting any two vertices is called a sector of the network. For a sector(I, j, d), the flow takes place in the direction $i \rightarrow j$. Accordingly, i is the initial (input) vertex and j the final (output) vertex of the sector considered.

Decision Variables

Define binary variable x_{ij}^d, with value 1 if the sector (I, j, d) is selected and 0 otherwise. To quantify the flow associated with each sector, we define the nonnegative continuous variable q_{ij}. We define the continuous variable p_{ij}^d to quantify the energy supplied in the region i of a sector (I, j, d).

Constraints

Initially, it must be ensured that all regions will be served by the network. Therefore, except for the destination region, each must have at least one sector through which to dispatch its production. Since the desired topology is a ST, the condition that only one flow exit exist per region shall be forced. Mathematically:

$$\sum_{j \in I, j \neq i} \sum_{d \in D} x_{ij}^d = 1, \forall i \in \overline{I},$$

(14)

That is, for every region $i \neq \theta$, only one pair (j, d) may be selected. It is important to note that the equation does not restrict the number of sectors ending at any region.

It is necessary to ensure the flow balance in the regions. For every region $i \neq \theta$, the incoming flow from $k \in \hat{i}$ at i, plus its own production, is sent to $j \in \hat{i}$, where $i \neq j \neq k$ to avoid cycles. Mathematically:

$$\sum_{j \in I, j \neq i} q_{ij} = Q_i + \sum_{k \in \overline{I}, k \neq i} q_{ki}, \forall i \in \overline{I}, j \neq k.$$

(15)

There is flow in sector (I, j, d) if it is selected, i.e., $qij > 0 \Leftrightarrow \exists d \in D : x_{ij}^d = 1$. To ensure that condition, the following constraints can be used:

$$Q_{MIN} \sum_{d \in D} x_{ij}^d \leq q_{ij} \leq \sum_{d \in D} Q_{MAX}^d x_{ij}^d,$$

$$\forall i \in \overline{I}, \forall j \in I, i \neq j,$$

(16)

where:

$$Q_{MIN} = \min_{i \in \overline{I}} \{Q_i\},$$

(17)

$$Q_{MAX}^d = \min \left\{ Q_{OPER}^d ; Q_T = \sum_{i \in \overline{I}} Q_i \right\}$$

(18)

The lower bound (Q_{MIN}) is defined as the lowest production of all regions. The upper bound (Q_{MAX}^d) is defined as the lesser value of the maximum flow supported by the pipe or the total output being drained along the network. That is, at no time can any flow traversing the sector (I, j, d) be below (Q_{MIN}) or above (Q_{MAX}^d), and flow is zero if the sector is not chosen.

No explicit constraint was introduced to ensure that the solution topology is an ST, or even to ensure flow to destination θ; nonetheless, restrictions (14), (15) and (16) together, ensure both conditions [18]. Basically, if the topology has cycles, the balance equations are satisfied only if there is more than one flow exit sector at some of the nodes of the cycle or if there is consumption at any of them.

Finally, to ensure energy balance, the amount of energy that must be provided to compensate for system losses must be estimated. In energy terms, each sector of the network can be viewed as an independent system with its own characteristics (flow, diameter, length, altitude difference etc.). Applying the assumptions of Section 3.2 to (2), it follows for every sector that:

$$p_{ij}^d = \left(\frac{8 f^d L_{ij} q_{ij}^2}{\pi^2 g \left(\phi^d \right)^5} + CP_{ij}^d - H_{ij} \right) x_{ij}^d, \forall i \in \overline{I},$$

$$\forall j \in I, \forall d \in D, i \neq j,$$

(19)

where CP_{ij}^d is the estimated losses due to geometry changes for the sector(i, j, d). By hypothesis,

$$p_{ij}^d = p_i - p_j = p_i$$
.

So P_{ij}^d represents the energy to be provided at i and not a pressure gradient within (i, j, d). Finally, the variable domain constraints:

$$0 \le x_{ij}^{d} \le 1, x_{ij}^{d} \in N, \forall i \in \overline{I}, \forall j \in I, \forall d \in D, i \ne j, \tag{20}$$

$$q_{ij} \in \Re_{+}, \forall i \in \overline{I}, \forall j \in I, i \ne j, \tag{21}$$

$$p_{ij}^{d} \in \Re_{+}, \forall i \in \overline{I}, \forall j \in I, \forall d \in D, i \ne j. \tag{22}$$

Objective Function

The Equation (13) will be taken as the objective function. Using the decision variables defined in Section 3.5.2, gives:

$$OF = \sum_{i \in \overline{I}} \sum_{j \in I, j \ne i} \sum_{d \in D} CC^{d} L_{ij} x_{ij}^{d}$$
$$+ \sum_{i \in \overline{I}} \sum_{j \in I, j \ne i} \sum_{d \in D} \left(CB_{ij}^{d} p_{ij}^{d} + CBC_{ij}^{d} x_{ij}^{d} \right), \tag{23}$$

which was generalized to consider all diameters in D. Note that when $x_{ij}^{d} = 0$ no contribution by any cost related to (i, j, d) is added to the project total cost.

General

The mathematical programming formulation (MF) of the problem is as follows:

MF: Minimize (23) subject to (14)-(16), (19)-(22).

SOLLUTION APPROACHES

General Considerations

MF can be manipulated. The p_{ij}^d variables appear only in the energy balance constraint (19), domain constraints (22) and objective function (23). The constraint (19) is an equality and, therefore, may be incorporated into the objective function. The variables p_{ij}^d can, in turn, be deleted from the formulation, together with constraints (22). Substituting (19) in (23), gives the new objective function of MF, where α^d and β_{ij}^d are appropriate parameters:

$$OF = \sum_{i \in \bar{I}} \sum_{j \in I, j \neq i} \sum_{d \in D} CC^d L_{ij} x_{ij}^d$$

$$+ \sum_{i \in \bar{I}} \sum_{j \in I, j \neq i} \sum_{d \in D} \left(CB_{ij}^d \left(\frac{\alpha^d L_{ij} q_{ij}^2}{(\phi)^{d5}} + \beta_{ij}^d \right) + CBC_{ij}^d \right) x_{ij}^d .$$

(24)

Heuristic

We consider (24) to develop the algorithm. By evaluating its parameters, three characterizations of a good solution to the problem can be inferred. First, consider the total length in a feasible solution S:

$$LT_{\Gamma(S)} = \sum_{(i,j) \in \Gamma(S)} L_{ij} ,$$

(25)

Where $\Gamma(S) = \{(i, j) x_{ij}^d = 1 \text{ for any } d \in D\}$. Both portions of cost in (24) are, in some sense, directly proportional to. $LT_{1(s)}$ The smaller the total length of the solution network, the lower the cost associated with it. Second, consider $\Gamma(S)$ (Figure 8), where v_0 is the destination vertex.

Additionally, suppose that v_6 is one of the vertices with highest output among producing regions and that, $L_{65} < L_{60} \ll L_{6 \to 0}$ where $L_{i \to j}$ represents the total length of the path connecting vertices i and j v L_{ij} i v

, and represents the direct connection between and j v . The following expression is computed for $_{G(S)}$ in (24):

$$\sum_{(i,j)\in\Gamma(S)}\left(\frac{L_{ij}q_{ij}^2}{\left(\phi^d\right)^5}\right),$$

(26)

That is, the cost is proportional to the square of the flow in (i j d) and the length of the sector. In addition, once flows increase in the direction of the destination, it can be advantageous to connect 6 to 0 v through the path that minimizes the distance between them (directly, for example) in order to reduce the cost of transportation (operation) between v_6 and v_0, in exchange for building $\Delta LT_{RD} = L_{60} - L_{65}$ additional units of pipe length. The Figure 9 shows the new solution G(S′)

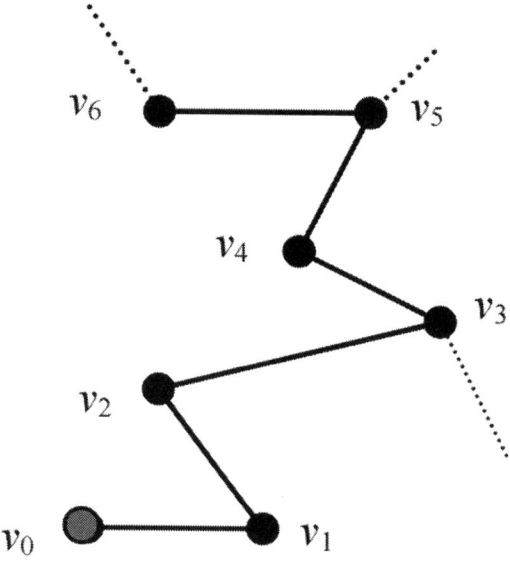

Figure 8: Final part of G(S) with destination v_0.

Finally, the cost of S is inversely proportional to the fifth power of the selected diameters and directly proportional to the flows transferred. Since flows increase in the direction of the destination, this increase can be offset by forcing the diameters at least not to decrease in the direction of flow. On the other hand, pipes of larger diameter are more expensive. Thus, it may be beneficial to increase the diameter of sectors with large flows, providing they are as short as possible.

Algorithm Description

Initially, the heuristic PND_H calculates the minimum spanning (MST) tree for the case being solved. This is because the MST is the best estimate in terms of total network length and may be used as initial search point for problems of this nature [19]. The algorithm then defines the flow direction and its value in each sector (i, j, d). The smallest available diameter capable of carrying the flow calculated for each sector is selected. The cost of the initial solution is stored. The algorithm attempts to decrease the distance traveled by large flows and increase the diameter of some sectors, one at a time—in both cases, a predefined number of times. At each trial (path reduction and diameter increase), the new solution is stored if it improves cost.

Exact

The MF is solved via an exact Outer Approximation algorithm proposed by Duran and Grossmann [20] for MINLP. Although the proposed model does not take into account all the necessary assumptions (linearity of the discrete variables, convexity of the nonlinear functions involving continuous variables and separability on the decision variables) to ensure the convergence of the algorithm to global optima, these global optima were obtained [12].

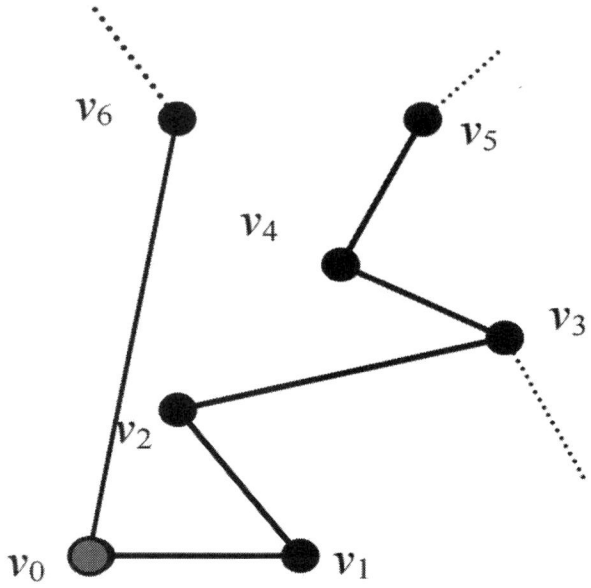

Figure 9: Final part of G(S') with destination v_0.

COMPUTATIONAL EXPERIMENTS

Computational experiments were conducted to test the efficiency and robustness of the model and algorithms. Because it is an original application to the PND problem, the literature on the subject contains no instances; these had first to be built. This section presents the cases constructed, the results obtained with the algorithms and the main conclusions from the analysis of these results.

Instances

To build instances, the sugar industry of São Paulo was taken as context. In particular, **Figure 10**shows the production capacity (millions of cubic meters) and its geographic distribution in 2006 [21]. The regions were named after the largest city in each region.

It can be readily seen how decentralized production is and how essential it is to use an appropriately designed and economically competitive logistics network to collect and concentrate production. Table 1 shows numerical values for output from each of the regions in 2006 [21].

Five instances were built, respectively with 4, 8, 12, 16 and 20 producing regions selected from among those listed in Table 1. Only five were built because, technically, a pipeline network designed to connect 20 producing regions can already be considered large. Besides, there was no access to real data.

The distances between regions were determined using the geographic coordinates of the main town. The height differences were calculated directly via the towns' altitudes. The outputs to be dispatched are listed in Table 1. To determine the associated flow of ethanol, the system was considered to operate three hundred days a year, twenty-four hours a day. The resulting flow rates are listed in Table 1. The idea is to consider network downtime indirectly.

To compose the instances, a set of six possible values for the pair "pipe diameter and thickness" was considered. Four of these pairs were associated with the smallest instance and all of them with the largest one. Table 2 shows the nominal values of the diameters and thicknesses:

Classical empirical equations were used to calculate the friction factors. There are three parameters to be considered in this calculation: diameter, relative roughness [14] and Reynolds Number. The diameters are listed in Table 2. The relative roughness is a characteristic of the pipeline construction material, and the material selected was carbon steel [9]. Reynolds Numbers cannot be calculated as a priori every pipe flow is unkown. Accordingly, an average flow traversing the network was considered for each available diameter.

In all cases, the loss estimator was considered similar for every(i, j, d), i.e., $CP_{ij}^d = CP$. Construction unit costs were determined using Equation (7). The unit cost of carbon steel was obtained in [22]. Since pump commercial prices and real consumption data (C_B costs) were not obtained, the methodologies proposed in 3.4.3 and 3.4.4 were not used. Thus, the coefficients of the second portion of (23) were sorted so as at least to respect the percentage given in Section 3.4. Table 3 summarizes the values of technical parameters used in constructing the instances. Table 4 summarizes the characteristics of the instances.

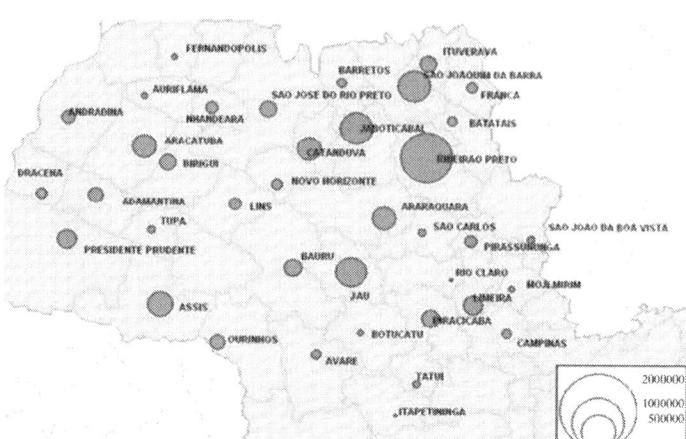

Figure 10: Geographic distribution of ethanol production in São Paulo—2006.

Table 1. Producer regions and associated output in São Paulo—2006

Producer region			
Name	**Identifier**	**Output (m³)**	**Flow (m³/h)**
Ribeirao Preto	RIB	1760574.84	244.524
Jaboticabal	JAB	838494.17	116.458
Sao Joaquim da Barra	SJB	827575.62	114.941
Jail	JAU	747952.94	103.882
Assis	ASS	539478.79	74.928
Catanduva	CAT	492136.98	68.352
Aracatuba	ARC	488406.16	67.834
Araraquara	ARR	472506.72	65.626
Limeira	LIM	333064.76	46.259
Piracicaba	PIR	322051.47	44.729

Presidente Prudente	PPR	296162.87	41.134
Bauru	BAU	287329.26	39.907
Ituverava	flu	253191.72	35.166
Adamantina	ADA	238679.21	33.150
Ourinhos	OUR	228766.52	31.773
Sao Jose do Rio Preto	SIR	209464.13	29.092
Birigui	BIR	207890.22	28.874
Andradina	AND	192191.56	26.693
Nhandeara	NHA	174069.49	24.176
Campinas	CAM	113655.93	15.786

Table 2: Pipe diameter and thickness values considered

Pipes		
Identifier	**Inner diameter (m)**	**Thickness (m)**
DO8	0.2032	0.0183
DIO	0.2540	0.0183
D12	0.3048	0.0214
DI4	0.3556	0.0238
DI6	0.4064	0.0262
DI8	0.4572	0.0294

Table 3: Adopted values to technical parameters

Parameter	Value (Unit)
Steel unit mass cost (M_c)	0.6 ($/kg)
Carbon steel density (r_M)	7860.0 (kg/m³)
Equivalent carbon steel roughness (M_e)	0.0045 (cm)
Safety factor (f)	0.72
Circumferential tension (s)	241.325 (MPa)
Ethanol density (g)	789.0 (kg/m³)
Ethanol viscosity (m)	1.19 e⁻³ (Ns/m²)
Head loss estimator (CP^a_{ij})	12.36 (m)
Gravity g	9.8 (m/s²)
Pi number (ϖ)	3.14

Table 4. Basic description of instances

Instance		
Name	Number of regions	Number of diameters
PND_SP_04_04	04	04
PND_SP_08_05	08	05
PND_SP_12_05	12	05
PND_SP_16_06	16	06
PND_SP_20_06	20	06

Results

The experiments were performed on an Intel Core 2 Duo 2.66 GHz with 8 GB of RAM. The PND_H heuristic was coded in ANSI C++. The MF was programmed in the commercial software AIMMS [23] and solved with the AOA package available in AIMMS. The OA algorithm problem solvers used were CPLEX 12.1 and CONOPT 3.14 M. The destination region is Campinas.

Table 5 shows the objective function values, PND_H and MF execution times, and a comparative analysis of results in terms of economic gains offered by the solutions. PND_H runtimes were null in all cases. MF produced the best solution for each case, which was obtained in low runtimes, as shown. By comparison, PND_H was able to find two optimal solutions (~0%). The others solutions entail small economic losses ranging from 2% to 6.7%.

We now analyze solutions in non-financial technical terms. Figure 11 represents the optimal design for the case PND_SP_04_04. It is topologically trivial. An interesting fact, however, is the reduction in diameter observed when passing through Limeira. We would expect the same diameter, since flow between Limeira and Campinas is larger. However, (5) suggests that a pipeline's maximum operating flow is inversely proportional to its length, which is shorter in sector Limeira-Campinas.

The PND_H solution to PND_SP_08_05 is shown in Figure 12. The network is 943.0 km in length and basically concentrates small outputs at Catanduva to be sent to Campinas via Jau. Only Ourinhos pipes directly to Jau, given their proximity. The production routed through Catanduva requires increased diameters on main sectors D12 and D14, successively. Figure 13 shows the MF solution. The network is 984.0 km length (41.0 km longer). Despite the increase, transferring output from Presidente Prudente to Jau via Ourinhos apparently reduces the transport cost (131.0 km shorter) and the construction cost of the Catanduva-Jau sector (from D12 to D10), since flow is decreased. The changes represent a1.72% saving in project cost (Table 5).

Table 5. Upper bounds generated by PNDH, optimal solutions obtained by MF, and percentage economic gains

PND_H			MF		
Instance	OF ($)	Time (s)	OF ($)	Time (s)	Economic gain (%)
PND_SP_04_04	43630086.61	0.0	43630086.61	0.00	0
PND_SP_08_05	96352887.34	0.0	94722203.45	0.05	1.72
PND_SP_12_05	118738486.56	0.0	118738486.50	0.23	$5.0e^{-08}$
PND_SP_16_06	148306176.40	0.0	146584104.40	0.33	1.17
PND_SP_20_06	185357307.04	0.0	173723124.10	37.56	6.7

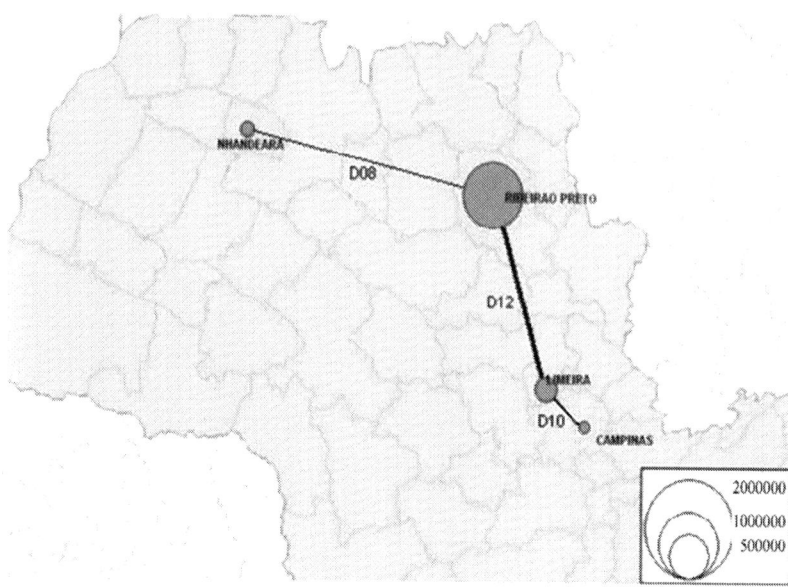

Figure 11: Optimal solution to instance PND_SP_04_04.

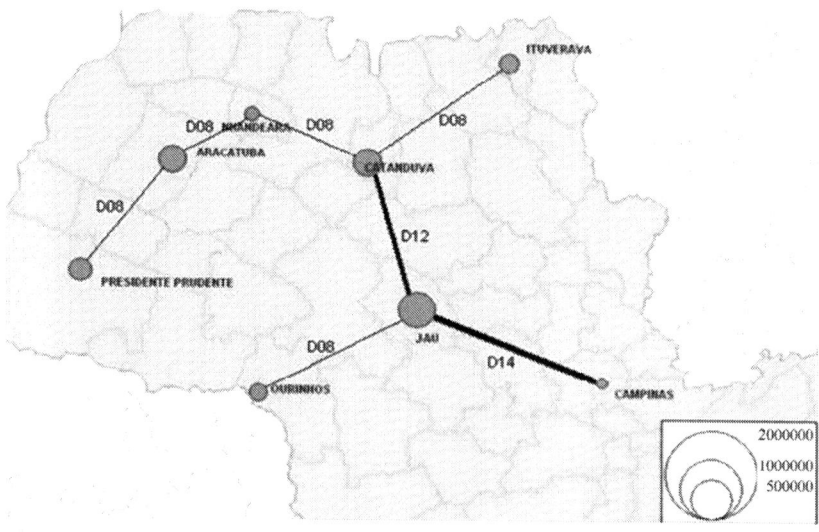

Figure 12. Solution to PND_SP_08_05 by PND$_H$.

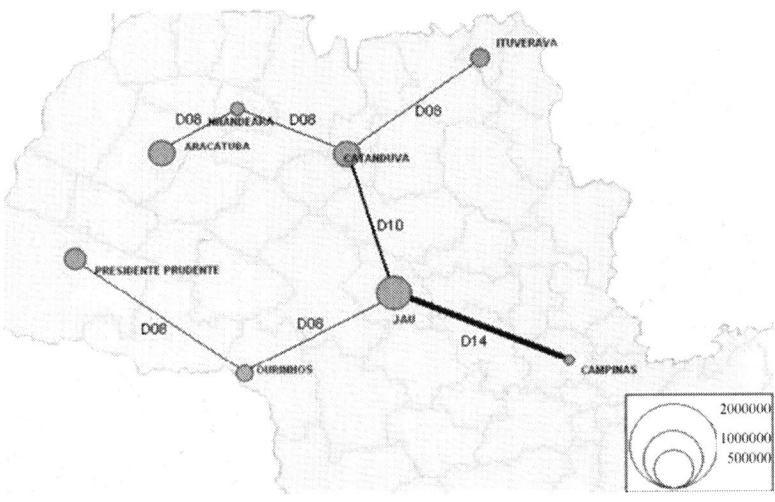

Figure 13. Optimal solution to instance PND_SP_08_05.

Figure 14 represents the optimal solution obtained to PND_SP_12_05, with 973.0 km total pipeline length. Interestingly, output from Nhandeara is initially transported to the west, away from Campinas, probably because the output is so small that it is not worthwhile connecting it to Ribeirao Preto or Bauru, 239.0 and 207.0 km away, respectively, even if the segment Adamantina-Assis (128.0 km) were removed, for example, in order to increase Nhandeara's throughput. Pipe diameter increases significantly at Jau, because it concentrates almost all output even before it arrives at Campinas. At Bauru, on the contrary, pipeline diameter is reduced, probably for reasons of geography.

PND$_H$ built the 1040.0 km solution to PND_SP_16_06, as sketched in Figure 15. Prioritizing the shortest possible pipeline length, Ribeirão Preto was connected to Jaboticabal (54.0 km northwest). Since Ribeirão Preto has the highest output (Table 1), pipeline diameter from Jaboticabal increased significantly (from D12 to D16). There is a diameter reduction at Limeira.

As compared with the optimal design (Figure 16), there is little change when leaving Ribeirão Preto, from which flow now travels 75.0 kilometers to Araraquara. This adds 21.0 km to the network (1061.0 km), reduces the construction costs of the Jaboticabal-Araraquara sector (from D16 to D12) and transport costs of output from Ribeirão Preto. The change represents a 1.17% project cost saving (Table 5). Finally, it is advantageous to transport output from Assis to Araraquara via Presidente Prudente (582.0 km) rather than connecting them directly. Increased production could justify a direct connection.

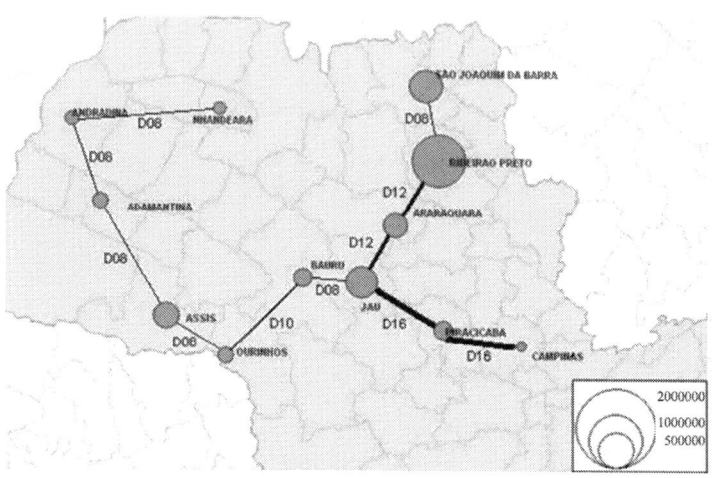

Figure 14. Optimal solution to instance PND_SP_12_05.

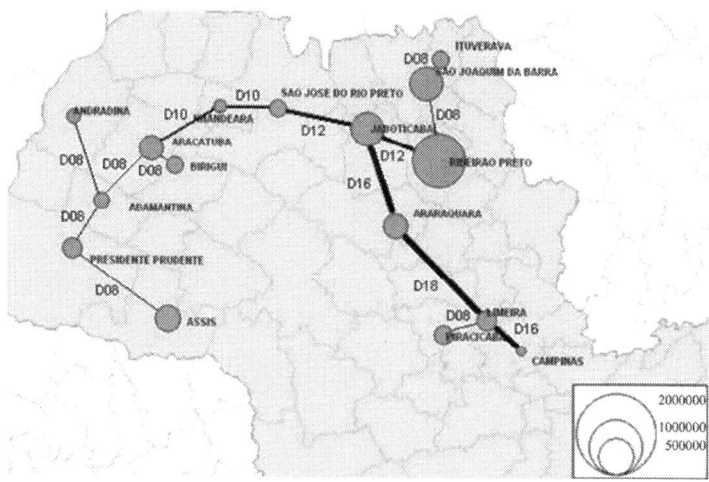

Figure 15. Solution to PND_SP_16_06 by PND$_H$.

Finally, Figure 17 represents the PND_H solution to PND_SP_20_06 (1361.0 km). Interestingly, Ribeirão Preto was connected to Campinas and not to a closer region (Araraquara or Limeira). This is because connecting it to Araraquara entails increasing diameters all the way to Campinas, thus also increasing the solution cost. Nevertheless, the Ribeirão Preto-Campinas connection does not form part of the optimal solution (Figure 18), since dispatching from Ribeirão Preto to Araraquara is economically favorable. This modification undermines the heuristic topology (Araraquara-Jau), making it necessary to detour flow from Araraquara to Limeira. This modifies pipe diameters in the final portion of the network, reduces the network length to 1282.0 km and represents a 6.7% saving in project cost (Table 5). Pipeline diameter is reduced at Piracicaba.

CONCLUSIONS

These findings emphasize the quality of the solutions built by PND_H (economic losses of less than 6.7% at null computational time cost). This result confirms the potential of the set of characterizations of a good solution. However, the most encouraging results were those generated by the exact algorithm. All except the largest case were solved in very low computational times. Improvement may possible be achieved by inputting information from the heuristic solution to the OA algorithm.

The tools developed to solve the PND addressed here met the expectations of robustness and efficiency and can be used as decision support tools for projects of this nature. Indeed, as the PND relates to a pipeline network whose purpose is to channel fluids to some central location, and where the main characteristic of the required solution is a topology without cycles and with pressure discontinuities at vertexes, these tools can be used.

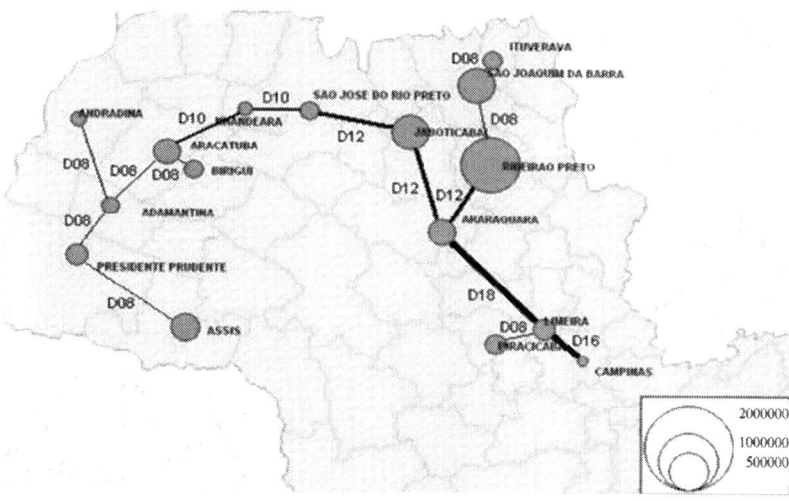

Figure 16. Optimal solution to instance PND_SP_16_06.

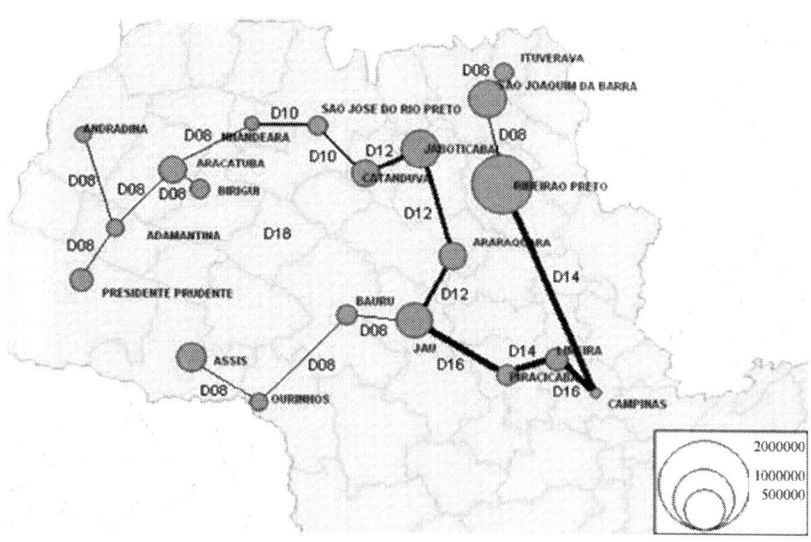

Figure 17. Solution to PND_SP_20_06 by PND$_H$.

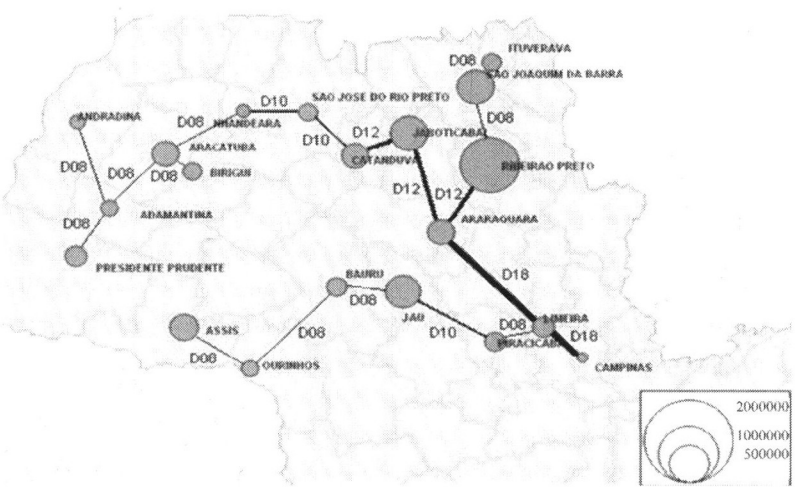

Figure 18. Optimal solution to instance PND_SP_20_06.

FINAL REMARKS

General

In this study, the problem of designing pipeline networks to transport fuel was addressed with a view to developing robust and efficient computational tools able to consider and assess the main design-related technical characteristics and decisions, as well as to estimate the construction cost and the operating cost (to any time horizon), in order to assist decision makers when evaluating the technical and economic feasibility of such projects.

Technical decisions are interdependent. For example: for a given flow rate, the larger the diameter of the duct, the lower the power dissipation. Thus pumps can be smaller and cheaper. However, the larger the diameter, the more expensive the pipe becomes. Conversely, if pipe diameter is diminished in order to reduce construction cost for the same flow rate, power dissipation is higher and thus higher-capacity (and more costly) pumps have to be installed. The larger the network, the harder it is to address these issues.

Accordingly, the combinatorial nature of the problem coupled with non-linear equations governing the flow energy balance, recommended the use of MINL mathematical programming models. To solve the problem, a heuristic algorithm that traces the main features of a technically feasible solution was developed to search for the best possible solutions in terms of cost. The mathematical program was coded in the AIMMS development environment, which provided the necessary optimization tools.

The particular context adopted was the ethanol industry in São Paulo State. The tools developed were tested on five cases built on the basis of output data from around the state. The computational results were extremely satisfactory, since all cases were solved efficiently by the heuristics, and also to global optimality by the exact algorithm, with low execution times, thus confirming the robustness of the approaches adopted.

Future Work

The most radical paradigm that can be broken is the deterministic approach. The assumption that the supply of ethanol is uncertain to the operating time horizon would probably lead to addressing the time variable directly, in order to design a network that best suits many different possible supply scenarios, as well as to using stochastic mathematical programs.

Technically, there are many options. Allocation of more than one diameter per sector of pipeline in the network may be considered. Pressure loss will be calculated as the sum of the losses in each of the sub-sections with different diameters. The energy balance equation should consider the different kinetic energy in the sectors, since different cross-sections produce different flow velocities for the same flow rate.

Another variant might consider finite storage capacities in the regions, to be decided in view of the related cost. This capacity is influenced by the stochasticity of production and the $NPSH_R$ pump parameter. If only the latter is considered, a stochastic approach to the problem can be avoided by estimating the minimum storage capacity of each region.

Finally, it would be useful to determine the number and location of pumps needed to power the system. Motivations to treat this decision are that it may not be possible to provide all the power necessary in each region by using one single station (technical limitation) and that, since pipes have maximum pressure tolerances, it may be necessary to distribute the energy supplied so that theses limits are observed.

ACKOWLEDGEMENTS

The authors would like to thank the Agência Nacional do Petróleo, Gás Natural e Biocombustíveis (ANP) for financial support provided under grant PRH-ANP-21.

REFERENCES

1. J. Goldemberg, "Ethanol for a Sustainable Energy Future," Science, Vol. 315, No. 5813, 2007, pp. 808-810. doi:10.1126/science.1137013

2. H. Pacini and S. Silveira, "Consumer Choice between Ethanol and Gasoline: Lessons from Brazil and Sweden," Energy Policy, Vol. 39, No. 11, 2010, pp. 6936-6942.doi:10.1016/j.enpol.2010.09.024

3. R. G. Compeán and K. R. Polenske, "Antagonistic Bioenergies: Technological Divergence of the Ethanol Industry in Brazil," Energy Policy, Vol. 39, No. 11, 2011, pp. 6951-6961.doi:10.1016/j.enpol.2010.11.017

4. Hira and L. G. Oliveira, "No Substitute for Oil? How Brazil Developed Its Ethanol Industry," Energy Policy, Vol. 37, No. 6, 2009, pp. 2450-2456.doi:10.1016/j.enpol.2009.02.037

5. T. Furtado, M. I. G. Scandiffio and L. A. B. Cortez, "The Brazilian Sugarcane Innovation System," Energy Policy, Vol. 39, No. 1, 2010, pp. 156-166.doi:10.1016/j.enpol.2010.09.023

6. União da Indústria de Cana de Açucar, "Dados e Cota- ções-Estatísticas," 2009. http://www.unica.com.br/dadosCotacao/estatistica

7. Centro de Estudos Logísticos e Instituto Brasileiro de Petróleo, Gás e Biocombstíveis, "Planejamento Integrado do Sistema Logístico de Distribuição de Combustíveis," 2008. www.ibp.org.br

8. L. C. Freitas and S. Kaneko, "Ethanol Demand in Brazil: Regional Approach," Energy Policy, Vol. 39, No. 5, 2011, pp. 2289-2298. doi:10.1016/j.enpol.2011.01.039

9. J. L. F. Freire, "Engenharia de Dutos," ABCM, Rio de Janeiro, 2009.

10. Project Management Institute, "Um Guia do Conjunto de Conhecimentos em Gerenciamento de Projetos (Guia PMBOK®)," 3rd Edition, PMI Publications, Newtown Square, 2004.

11. M. Minoux, "Network Synthesis and Optimum Network Design Problems: Models, Solution Methods and Applications," Networks, Vol. 19, No. 3, 1989, pp. 313-360.doi:10.1002/net.3230190305

12. G. D. Silva, "Abordagens Heurísticas e Exatas para o Problema de Projeto de Redes de Dutos," M.Sc. Thesis, COPPE/UFRJ, Rio de Janeiro, 2009.

13. J. Brimberg, P. Hansen and K. Lih, "An Oil Pipeline Design Problem," Operations Research, Vol. 51, No. 2, 2003, pp. 228-239. doi:10.1287/opre.51.2.228.12786

14. Munson, D. F. Young and T. H. Okiishi, "Fundamentos da Mecânica dos Fluidos," 4th Edition, Edgard Blücher, São Paulo, 2004.

15. J. R. Moreira and J. Goldemberg, "The Alcohol Program," Energy Policy, Vol. 27, No. 4, 1999, pp. 229-245. doi:10.1016/S0301-4215(99)00005-1

16. Forlano and R. Maclain, "Facilities Piping," Paragon Engineering Services, Houston, 1991.

17. F. Harary, "Graphs Theory," Addison-Wesley, Reading, 1969.

18. B. Gavish, "Topological Design of Centralized Computer Networks—Formulations and Algorithms," Networks, Vol. 12, No. 4, 1982, pp. 355-377. doi:10.1002/net.3230120402

19. F. Rothlauf, "On Optimal Solutions for the Optimal Communication Spanning Tree Problem," Operations Research, Vol. 57, No. 2, 2009, pp. 413-425.doi:10.1287/opre.1080.0592

20. M. A. Duran and I. E. Grossmann, "An Outer-Approximation Algorithm for a Class of Mixed-Integer Nonlinear Programs," Mathematical Programming, Vol. 36, No. 3, 1986, pp. 307-339. doi:10.1007/BF02592064

21. L. N. S. M. Dória, "Construção e Entendimento dos Cenários dos Mercados Interno e Externo do Álcool no Brasil (2006-2015)," Final paper on course in Operation Management in Oil Exploration and Production, Polytechnic School, Rio de Janeiro, 2007.

22. MEPS LTD, "MEPS—World Carbon Steel Prices—With Individual Product Forecasts," 2009. http://www.meps.co.uk/World%20 Carbon%20Price.htm

23. M. Roelofs and J. Bisschop, "AIMMS User's Guide," Paragon Decision Technology B.V., 2011. http://www.aimms.com/ downloads/manuals/user-s-guide

On the Flow Characteristics of the Conical Minoan Pipes Used in Water Supply Systems, Via Computational Fluid Dynamics Simulations

G. Tseropoulos[a], Y. Dimakopoulos[a],
J. Tsamopoulos[a], and G. Lyberatos[b, c]

[a]Laboratory of Computational Fluid Dynamics, Department of Chemical Engineering, University of Patras, Karatheodori 1, 26500 Patras, Greece

[b]School of Chemical Engineering, National Technical University of Athens, Greece

[c]Institute of Chemical Engineering & High Temperature Chemical Processes, ICE-HT/FORTH, Greece

ABSTRACT

The Minoan Terracotta pipes with their conical shape were widely used in the water distribution system in the ancient Minoan civilization.

They remain one of the brightest achievements of the Minoan tribe in water supply technology and raise admiration as well as many questions about the technological advancements of antiquity, that are yet to be understood. The present work aims at answering the following questions: a) what inspired the Minoans to manufacture pipes with such a peculiar shape, that differs greatly not only from later pipe designs of antiquity, but also from contemporary cylindrical pipes and b) why was the design of those pipes abandoned after the fall of the Minoan civilization? It tries to address these questions by investigating the flow physics and dynamics that take place in such pipes, adopting advanced numerical and computational methods. The time-averaged Navier–Stokes equations along with the k – ε turbulence model are solved for a variety of geometrical parameters, pipe orientations and flow rates, in order to produce a comparative picture of the hydraulic efficiency of the conical Minoan pipes. The flow field is visualized and critical flow parameters, such as the head loss, the velocity magnitude and turbulence intensity, are calculated. These calculations show clearly that the conical Minoan pipes exhibit significantly higher pressure drops along their length compared to an equivalent straight pipe. In their widest part an extended recirculation appears, which could wash out impurities that may be present in the water, which at the same time cannot be deposited on the pipe wall. This evidence proves that the Minoan pipes are energetically expensive to operate and consequently their replacing by cylindrical pipes was inevitable. Therefore, it seems that the main advantage and purpose of the particular geometry was that they could be easily connected on site, forming long straight or slowly bending lines without having to add straight or many different fittings in between.

INTRODUCTION

The Minoan culture flourished during the Bronze age and achieved a variety of advances in science and technology, especially in water management. This was quite important, because water resources were not abundant in their area. A variety of technologies such as rainwater harvesting, wells, cisterns, gutters, and sedimentation tanks as well as flat rooftops and light wells were used to facilitate the water management starting in the Early Minoan period, ca. 3500–2150 BC

(Mays, 2010; Angelakis and Spyridakis, 2010;Angelakis et al., 2005). Although the Minoans mainly relied for their water needs on wells that pierced through the aquifer, the transportation of water over long distances through pipes and channels was well developed, due to the mountainous terrain of the island of Crete. Open stone conduits with U-shaped cross-section were used to transfer water from stone made sedimentation tanks to the main storage cistern. On the other hand, urban water distribution to the palace and the settlements used closed type pipes of terracotta, that were found in abundance in the major cities of Crete including Knossos, while others survive in bad condition at the palace of Phestos and other places in Crete. Unlike pipes elsewhere in Greece of later period, Minoan pipes were manufactured in conical sections with their ends shaped properly in order to provide neatly fitting joints (Angelakis et al., 2007; Webster and Hughes, 2010). This closed system implies a practical knowledge of engineering and physics, as the design of these pipelines has taken into account many hydraulic principles as well as the principle of communicating vessels (Fig. 1 and Fig. 2).

Figure 1: Clay pipe from Knossos.

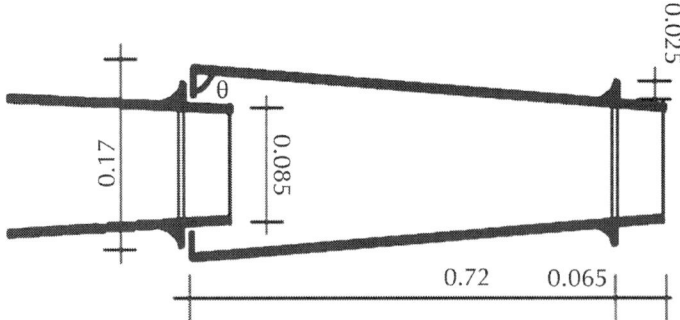

Figure 2: Typical dimensions in meters of a Terracotta pipe used for water supply in Knossos (the figure is not in scale).

Scientific proof concerning the achievements mentioned above was presented by Arthur Evans (1921–1935) when he discovered that the Minoans used the latter principle for the purpose of transferring water from distant rivers into inhabited regions. As an example he mentions transferring water from the river of Maurokalamos to the Minoan palace in the ancient city of Iraklio, where nowadays many remains of internal installations of sewerage and drainage systems are to be seen. Such was the competence of the Minoan craftsmen that these systems are operational even today, 5000 years after their construction. In order to transfer water in the residences, closed (or sometimes open) ceramic pipes were used, which perfectly fitted into each other, in order to avoid clogging or leakage phenomena. Those pipes were manufactured with variable diameters, presumably so that the pressure that caused the water flow, would not allow any solids accumulation inside the pipe (Dialynas et al., 2006). These pipe segments made by Terracotta indicate that the Minoans were also very concerned about sanitation, as they used (unclean) stone pipes more often for sewage waters that were driven away from the residences (Mays, 2008).

The technological advances of the Minoan civilization and their adequate knowledge of the principles of physics has led current researchers to believe that a deeper reason, that involves hydraulic phenomena might had led them to the construction of the conical pipes mentioned above instead of cylindrical ones (Webster and Hughes, 2010). If this is the case, it becomes unclear why the design of those pipes was abandoned after the fall of the Minoan civilization and what drove the Minoans in manufacturing such a peculiar pipe, that differs greatly not only from later pipe designs of antiquity, but also from contemporary cylindrical pipes. Consequently, significant issues related with the hydraulic efficiency and the shape of the tube, should be addressed in order to safely explain all the stages in the advancement of the water supply pipe technology in the Minoan civilization or at least to exclude some of the prior hypotheses on this matter. Clearly, during the Minoan era, aspects of hydraulics knowledge came from empirical observations translated into practical working principles. However, modern fluid mechanics helps us avoid this empiricism by employing either experiments or models. The latter can be solved either analytically after certain simplifying assumptions and approximations or numerically, which usually provide more accurate predictions at high Reynolds number flows where turbulence is the prevailing mechanism for momentum transfer and controls the scales of the flow. Also these numerical solutions allow us to visualize the flow without performing experiments, which can be more difficult to control and more time-consuming. We also give approximate analytical solutions based on semi-empirical expressions to obtain results and answers without requiring the sometimes costly numerical simulations. The results under the same conditions should be comparable and provide a nice test for both approaches.

Computational Fluid Dynamics (CFD) is not only a way to simulate industrial processes or physical problems, but also an everyday tool to deal with common or complex problems that include simple or complex geometries. Typical applications of the CFD techniques in the field of engineering are the coating flows over topographies (Pavlidis et al., 2010; Papaioannou et al., 2009), polymer processing flows (Dimakopoulos and Tsamopoulos, 2006), the two phase flow of oil and water in porous media (Kouris et al., 2002) etc. On the other hand, CFD has been more rarely used and in a less systematic manner in the science of archaeology. Through CFD models scientists can simulate the flow patterns inside archaeological findings that involve technological achievements of ancient civilizations, predict their overall behavior, explain their utility and even give answers to questions that appear obscure due to the lack of archaeological data. Indicative studies are: the determination of the flow inside of the pipelines of Aspendos city (Nicolic, 2008), the investigation of the hydraulics of a roman siphon in Aspendos city (Ortloff and Kassinos, 2003; Ball, 2003), the simulation of the urban water supply and distribution system (Ortloff and Crouch, 2001) in Ephesos city, and the study of the hydraulic performance of ancient waterwheels (Pujol and Montoro, 2010). Furthermore, recent works show increasing usage of CFD for the purpose of analyzing, interpreting, developing and improving technological achievements of ancient civilizations. Using theoretical knowledge of control systems and combining it with CFD, the water supply system of city of Gaziantep in Turkey has been simulated and analyzed in order to represent the system of pipelines, reservoirs and water pumps responsible for the water supply of the city and compare it with real-time measured data (Eker et al., 2002). Extensive analysis of the supply system of the city of Apamea (Syria) has provided a qualitative description of the water supply system of the city, using CFD simulations in order to compute several data such as the water flow rate and the energy loses (Haut and Viviers, 2007).

The remainder of this article is organized as follows. The physical problem as well as the governing equations and the turbulence model adopted in this work are presented in the next section. The results are presented and discussed in Section 3, while final conclusions are drawn in Section 4.

PHYSICAL MODEL

We consider the flow of water inside a Minoan pipe under steady conditions, which is caused by a constant pressure gradient (Fig. 3). This pressure gradient generates a constant flow rate typically of Q ≈ 8 l/s, and forces the flow of water from the left to the right. The entire pipe is composed of many repeated units (segments) like the one shown in Fig. 3. In the entire pipe, the pressure gradient and the flow rate remain the same as in the segment examined in this study. The exit cross section of a segment of the Minoan pipe is smaller and is used to direct water to the next segment. There, the cross section undergoes an abrupt expansion. The two consecutive pipe segments are tightly fit to avoid leakage. The flow domain in a pipe segment is assumed to be symmetric with respect to the vertical xy plane and periodic in the x direction. Although the pipe design is axisymmetric, no axisymmetry in the flow field can be assumed because of the gravity force in the y direction and the distorted turbulent eddies arising in the flow. The dimensions of the computational domain are summarized in Table 1. Water typically exhibits only viscous behavior and has constant viscosity and density. Therefore, its flow is incompressible. Its properties at 25 °C are presented inTable 2.

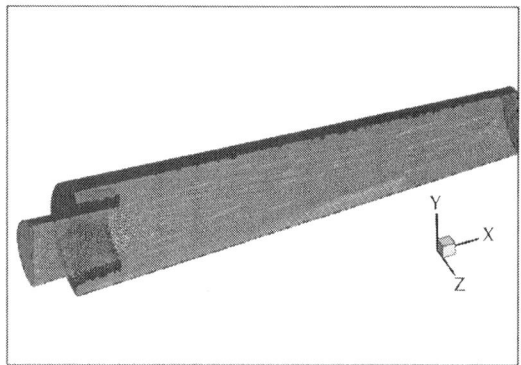

Figure 3: Mesh used to compute the flow field in the Minoan pipe.

Table 1: Typical geometrical dimensions of Minoan pipes

Parameter name	Value	Units
Diameter of the inlet section	0.17	m
Diameter of the outlet section	0.085	m
Ratio of the outlet section to the inlet section diameters, Y	1/2	–
Total length	0.72	m

Table 2: Physical properties of water and flow parameters

Parameter name	Value	Units
Ambient pressure	1 (101.3)	Atm (kPa)
Ambient temperature	25	°C
Density, ρ	998	kg/m3
Viscosity, μ	0.001	kg/m-s
Basic flow rate, Q	8	l/s

The water flow is governed by the momentum and mass conservation equations, which in their dimensional form are:

$$\rho \left(\frac{\partial u_i}{\partial t} + u_j \frac{\partial u_i}{\partial x_j} \right) = -\frac{\partial P}{\partial x_i} + \frac{\partial}{\partial x_j} \left((\mu + \mu_t) \frac{\partial u_i}{\partial x_j} \right) + \rho g_i$$

(1)

$$\frac{\partial u_i}{\partial x_i} = 0$$

(2)

where u_i is the component of the velocity vector in the x_i direction, P is the pressure, ρ is the density, μ is the molecular viscosity, μ_t is the turbulent (eddy) viscosity and is related to the so-called Reynolds stresses, g_i is the component of gravity in the x_i direction. A repeated index in a term of the above equations indicates an inner product of the corresponding vectors.

Since we have assumed that the flow field is periodic, pressure can be decomposed into two parts the periodic component P_p and the gradient of a linearly varying component in the flow direction, $(\Delta P/L)x$, which controls the flow rate:

$$P \equiv P_p + \frac{\Delta P}{L}x$$

(3)

For the physical parameters of water (Table 2), the value of the Reynolds number, Re, based on the mean velocity $u_m \equiv 4Q/(\varpi D^2)$ and the tube diameter, D at the inlet, exceeds the critical value for laminar flow, which is ~2300

$$Re \equiv \frac{\rho u_m D}{\mu} = 59,950$$

(4)

Therefore the flow is turbulent and tiny eddies arise in it. In order to capture such small scale flows, extremely fine meshes are required, when Direct Numerical Simulation (DNS) techniques are applied. This makes the DNS method very expensive in terms of memory and processor requirements. Instead, we use the Reynolds-Averaged Navier–Stokes (RANS) equations (Yih, 1977; Deen, 1998). Their derivation is based on assuming that the velocity field $\underline{u}$ is decomposed in a time-averaged component $\underline{U}$ and a component $\underline{u}'$, which is quickly varying in time. Then, in the averaged momentum balances the Reynolds stresses arise, which are composed of averages of products of the velocity fluctuations, $R_{ij} = u'_i u'_j$. In order to calculate them, we apply the k–ε model. This model includes two equations, additional to the momentum balance, to compute the turbulent properties of the flow, while it takes into consideration effects related to convection and diffusion of the turbulent energy. The first equation describes the kinetic energy, k, and the second one the turbulence dissipation rate, ε. In k–ε models μ_t is related to k and ε. The specific parameters of the model and their values, taken from (Poroseva, 2001) assuming perfectly smooth pipe walls, are given in Table 3. As far as the boundary conditions are concerned, we impose that all variables ($\underline{U}$, P, k, ε) are periodic between the inlet and outlet cross-sections (red-colored surfaces in Fig. 3 (in the web version)), symmetric along the xy plane to reduce the size of the problem (green-colored surface in Fig. 3). On the pipe wall (blue-colored surfaces in Fig. 3) we impose no-slip and no-penetration conditions for the velocity field and the k–ε model variable.

Table 3: Turbulence constants in the equations for the k–ε model for a pipe with smooth walls

Dimensionless parameter name	Value
σk	1.0
σ	1.3
C1	1.44
C2	1.92
Cμ	0.09

The equations governing the flow are non-linear and cannot be solved analytically. Therefore, their solution can be obtained only by numerical methods. Because of the assumption of symmetry at the mid-plane, the domain in which we solve these equations numerically is half the meshed volume of the pipe. This domain is tessellated, i.e. it is divided into a set of small tetrahedral or hexahedral cells. First, we set the number of elements along the pipe wall and in the radial direction in the exit of the pipe, Then, the software "GAMBIT" (part of the FLUENT package) completes the discretization of the volume inside the pipe, paying special attention in resolving the flow near the boundaries. After the usual convergence studies with mesh refinement in our simulations, we found out that approximately 200,000 elements (cells) are needed for the converging pipes and 400,000 elements for the straight pipes, because of their larger volume. Having set the convergence criterion (relative Euclidean norm of the residuals) for the iterations at 10^{-7}–10^{-6}, the majority of the computational experiments with the final mesh converged after up to ~2000 iterations. Specific data about the number of cells that were used, the number of elements along the pipe wall, the number of elements in the radial direction at the pipe outlet, the number of iterations to achieve convergence and the computation times are given in Table 4. The resulting length of each side of an element is less than 3 mm. The conservation equations are integrated and applied to each of these discretization cells. Through this method, we obtain a system of non-linear algebraic equations that is solved numerically with a method that is pressure-based and Green-Gauss cell-based as a Gradient option. The discretization of the differential equations using the finite volume technique and the numerical solution of the resulting algebraic ones are performed using the commercial software "FLUENT" (Fluent™, ANSYS, 2012).

Table 4: Characteristics of numerical calculations. In the last two columns we give the number of elements along the length of the pipe wall and along the pipe diameter at its exit, respectively

Geometry	Number of elements	Computation time (sec)	Number of iterations	Elements x-direction	Elements Doutlet
Straight	399,205	18,139	1035	260	62

Conical Pipe (Y = 2/3)	209,678	23,471	1271	259	43
Minoan Pipe (Y = 1/2)	195,729	24,196	1458	259	29
Conical Pipe (Y = 1/4)	185,975	26,712	1924	259	15

RESULTS & DISCUSSION

The questions that were posed in the beginning of this study were to determine the reasons that drove the Minoans to design conical pipes instead of cylindrical ones and why this particular design was abandoned in later times. To answer these questions we designed three different numerical experiments, which enabled us to see the operation of the Minoan pipe in comparison with other designs or under different flow conditions. In the first study four different geometrical cases are compared: the straight pipe, the original Minoan pipe design having a diameter ratio of the narrow to the wide cross section, Y, equal to 1/2, and two additional designs having Y = 2/3 and 1/4. The entrance of the pipe is always its widest part and the radius there is kept constant between the different geometries. The second study examines the flow structures in the Minoan tube at different flow rates, and the third study presents how the efficiency of the Minoan design varies with the orientation of the pipe with respect to gravity. The results are shown below, in the form of graphs. Subsequently, the pressure drop along each pipe section, together with the drag force on the wall are calculated and given in Table 5 and Table 6. The drag force is defined as the force component exerted from the fluid on the pipe wall and in the direction of the flow velocity. It is given by the expression known as drag equation:

$$F_D \equiv \left(\underline{\underline{\tau}} \cdot \underline{n} \right) \cdot \underline{t} \ = \ (\mu + \mu_t) \left[\left(\nabla \underline{U} + \overset{+}{\nabla \underline{U}} \right) \cdot \underline{n} \right] \cdot \underline{t}$$

$$(5)$$

where F_D is the drag force, $\underline{\underline{\tau}}$ is the stress tensor, which is defined in Eq. (5) and $\underline{t}$ and $\underline{n}$ are the tangent and normal vectors on the pipe wall.

Effect of Geometry on Flow Characteristics

In order to visualize the effect of the pipe geometry on the flow field, first we plot the calculated streamlines for the four geometries at the xy symmetry plane (Fig. 4). Under steady state conditions the streamlines represent the time-averaged trajectories of the fluid elements, which remain on this plane due to symmetry. In the case of a straight pipe they are straight lines, which are parallel to the tube wall. When the diameter ratio Y decreases, closed loops of streamlines arise corresponding to a large recirculation and covering an annular part of the pipe inner space near the entrance. Flow recirculation arises because of the gradual increase of the average velocity as the cross sections get smaller closer to the end of the pipe and its abrupt decrease downstream when the cross section area abruptly increases. Hence, naturally their size becomes larger as the inner radius at the entrance section gets smaller. This recirculation contributes into stirring the material near the pipe wall and, thus, into cleaning the wall from possible deposits. However, behind it and in the region with the largest diameter, the fluid hardly moves and any particles in the water may accumulate there. The fluid entrapped in a recirculation will not move downstream via the mean velocity, but via the velocity fluctuations caused by turbulence. The recirculation is not symmetric with respect to xz plane because of gravity. In the core section of the tube, open streamlines are seen, which are nearly straight, because the fluid is strongly convected in the x-direction (Fig. 4).

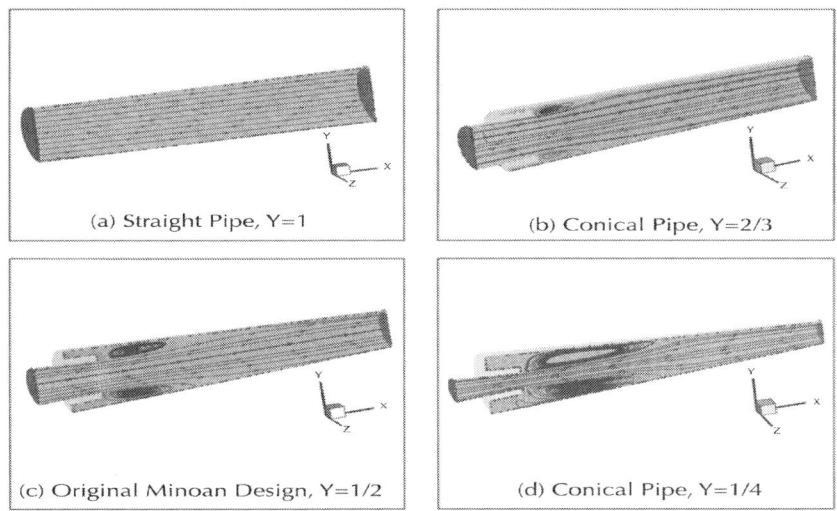

(a) Straight Pipe, Y=1

(b) Conical Pipe, Y=2/3

(c) Original Minoan Design, Y=1/2

(d) Conical Pipe, Y=1/4

Figure 4: Effect of pipe geometry on the streamlines under the same flow conditions (Table 2) a) Straight pipe, Y = 1, b) Conical pipe withY = 2/3, c) Original Minoan design, Y = 1/2, d) Conical pipe with Y = 1/4.

The contours of the velocity magnitude ($|\underline{U}|$) are given in Fig. 5, also at the xy symmetry plane. They verify the above mentioned observations and simultaneously offer a quantitative picture of the effect of the radii ratio, Y on the flow. The maximum time-averaged velocity appears always in the core region of the tube, just after the entrance section. For a straight pipe, Y = 1, the average value is ~0.35 m/s, as it should, given that Q ≈ 8 l/s and D = 0.17 m, and remains nearly constant along the pipe. For a Minoan pipe with Y = 1/2 the maximum velocity arises inside the entrance and before the abrupt expansion of the cross section. In fact, it increases in proportion to the square of the radii ratio i.e. it is 4 times higher than in the straight pipe, and it is equal to ~1.4 m/s, simply because of mass conservation.

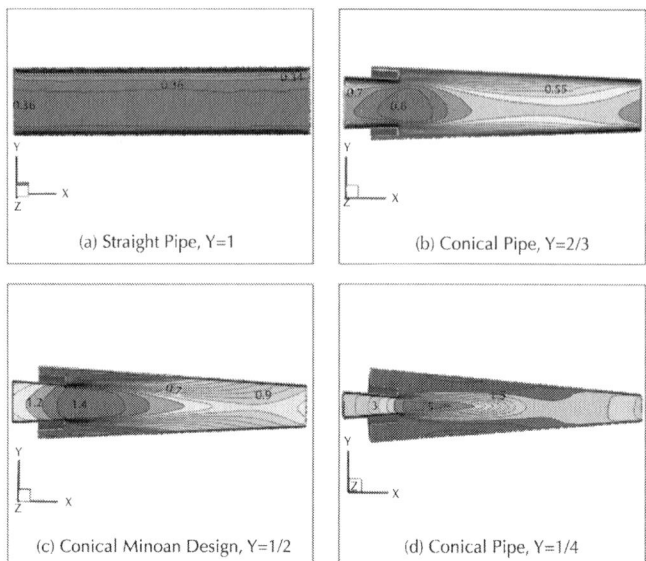

Figure 5: Effect of pipe geometry on the contours of the mean velocity magnitude ($|\underline{U}|$) in [m/s] for four different designs (for Q = 8 l/s): a) Straight pipe, Y = 1, b) Conical pipe with Y = 2/3, c) Original Minoan design, Y = 1/2, and d) Conical pipe with Y = 1/4.

Similar is the effect of Y on the turbulence intensity, which is given at the xy symmetry plane as well (Fig. 6). The turbulence intensity is a relative turbulence scale, which is defined as the ratio of the measure of the velocity fluctuations with respect to the mean velocity, $|\underline{u}'|/|\underline{U}|$. The quantities $|\underline{u}'|$, $|\underline{U}|$ are given by the following expressions.

$$\left|\underline{u}'\right| = \sqrt{\frac{1}{3}\left(u_x'^2 + u_y'^2 + u_z'^2\right)} = \sqrt{\frac{2}{3}k}$$

(6)

$$\left|\underline{U}\right| = \sqrt{U_x^2 + U_y^2 + U_z^2}$$

(7)

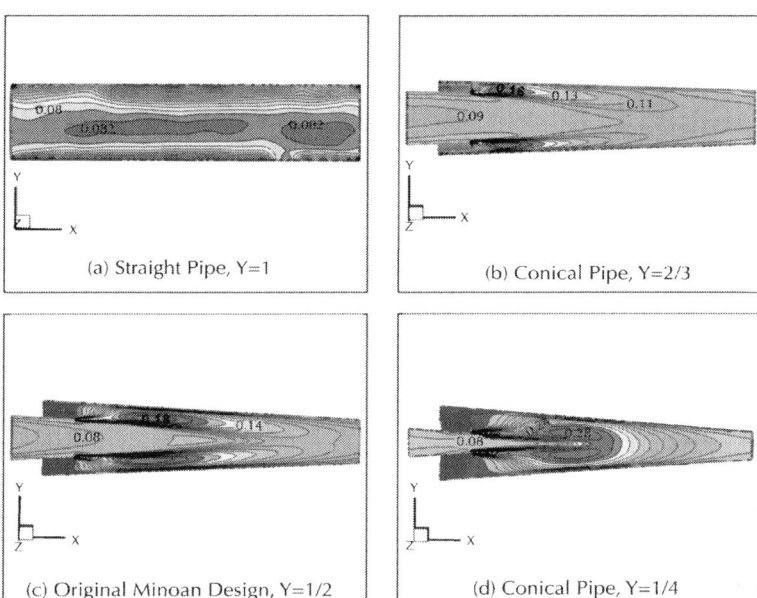

Figure 6: Effect of pipe geometry on the contours of the turbulence intensity (u'/U) for four different designs (for Q = 8 l/s): a) Straight pipe, Y = 1, b) Conical pipe with Y = 2/3, c) Original Minoan design, Y = 1/2, d) Conical pipe with Y = 1/4.

Values of $|\underline{u}'|/|\underline{U}|$ up to 1% indicate low turbulence flows, values between 1% and 5% indicate medium turbulence flows and finally values higher than 5% correspond to high turbulence regimes. In Fig. 6 we can observe that the flow in a straight pipe, Y = 1, with flow rate of Q ≈ 8 l/s has a uniform turbulence intensity of about 8%. Turbulence intensity increases as Y decreases so that its maximum value for Y = 1/4 is 28%. Even in the Minoan pipe, the flow is highly turbulent with intensity values of ~8% in the pipe entrance and ~18% in the area of the large recirculation eddies. Its maximum value is ~2.5 times higher than the max value of the straight geometry.

In the case of conical pipes, the region where turbulence intensity is maximized coincides with the region where the strain rate, $\nabla \underline{U} + \nabla \underline{U}^+$, is maximized (Fig. 7). These regions appear around the entrance section of the tube, where sharp boundary layers and recirculation develop (see also Fig. 4). The boundary layers are separated from the pipe wall and convected by the high velocity stream of water in the core region. Decreasing Y, decreases the angle ϑ, which is formed by the inlet surface of the conical segment and the wall attached to it. Angle θ varies from 90° (straight pipe) to 84.85° (Y = 1/4). Its values are also given in Table 5.

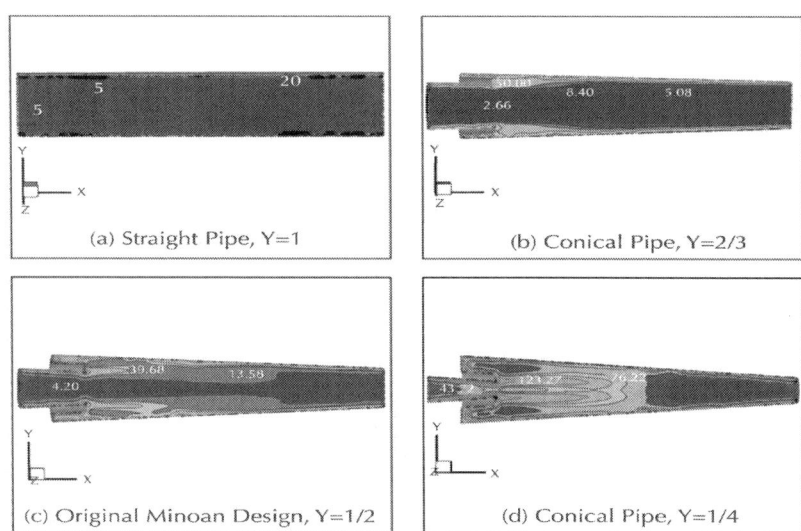

Figure 7: Effect of pipe geometry on the contours of the strain rate in [s⁻¹] for four different designs (for Q = 8 l/s): a) Straight pipe, Y = 1, b) Conical pipe with Y = 2/3, c) Original Minoan design, Y = 1/2, d) Conical pipe with Y = 1/4.

Table 5: Effect of ratio of min to max diameter, $Y = D_{min}/D_{max}$, on pressure drop and drag coefficient for Q = 8 l/s (the radius of the inlet section is kept constant as Y decreases for the different conical geometries)

Y (angle θ)	ΔP/L (Pa/m)	FD (N)
1(90°)	−21.089	0.151
2/3(87.69°)	−172.897	1.498
1/2(86.56°)	−724.982	3.406
1/4(84.85°)	−1652.246	6.175

Considering the findings from Fig. 4, Fig. 6 and Fig. 7, we can safely conclude that the Minoan pipe exhibits high losses of energy due to turbulence dissipation and the recirculation in the annular section of the tube. We can determine the loss of energy by calculating the pressure drop $\Delta P/L$ and the drag force. Their values are given in Table 5 and demonstrate a rise of the pressure drop as well as of the drag force when the parameter Y decreases. In the straight pipe both $\Delta P/L$ and F_D are relatively small. Indicatively, the ratio of pressure drop in the Minoan pipe to that of a straight one is almost 35 for a flow rate of $Q \approx 8$ l/s. It is noteworthy that small decreases in the inclination angle θ of the tube wall cause very large increases in the turbulence intensity and energy loses.

We can compare these numerical predictions with results from approximate correlations and theory, which are given in standard Fluid Mechanics books (e.g. Çengel and Cimbala, 2010). For example, the pressure drop per pipe length $\Delta P/L$ can be computed using the expression

$$\frac{\Delta P}{L} = \frac{f}{D} \frac{\rho u_m^2}{2} = \frac{f}{D} \frac{\rho}{2} \left(\frac{4Q}{\pi D^2} \right)^2$$

(8)

Where f is the empirical friction factor and the second fraction in the right-hand-side is the kinetic energy per unit volume at a cross section of the pipe. The friction factor can be obtained by various correlations such as those proposed by Colebrook or Darcy–Weisbach. Here we will use the correlation by Haaland (1983), which is equally accurate, but simpler to use because it provides an explicit expression for f:

$$\frac{1}{\sqrt{f}} = -1.8 \log \left[\frac{6.9}{Re} + \left(\frac{\varepsilon/D}{3.7} \right)^{1.11} \right]$$

(9)

where ε is the "roughness" of the pipe wall. The actual roughness of the pipes while they were used in the Minoan era is not known, but certainly it is expected to be rather high. Hence the energy loss due to friction during their operation would be higher than for smooth pipes, which we have assumed for the numerical predictions above. With these approximate expressions we will be able to find the effect of the surface roughness on the energy loss. First, let us assume that the pipe is perfectly smooth, $\varepsilon = 0$. Then for $Q \approx 8$ l/s Eq. (9) yields $f \sim 0.02$ and Eq. (8) predicts $\Delta P/L = 7.3$ Pa/m, which is considerably lower than the value 21.1 given in Table 5 for a straight pipe. Indicatively, if the pipe roughness increases to $\varepsilon = 0.0002$ m or 0.001 m, then $f = 0.024$ or 0.034 and $\Delta P/L = 8.8$ Pa/m or 12.4 Pa/m, i.e. the energy loss increases, as expected. If we examine the Minoan pipe or the other converging pipes, we may assume that most of the energy loss is caused by the converging walls and the sudden expansion in the entrance of the pipe, ignoring as subdominant the energy loss by the wall roughness or the usual friction in homogeneous turbulence, which we studied above numerically. Then the pressure drop per pipe length $\Delta P/L$ can be computed using the equation:

$$\frac{\Delta P}{L} \equiv \left(k_{exp} + k_{conv}\right) \frac{\rho u_m^2}{2L}$$

(10)

where k_{exp} and k_{conv} are the energy loss coefficients for the sudden expansion and wall convergence. Typical values and ways to calculate them are reported in Çengel and Cimbala (2010). For the Minoan pipe we find $k_{exp} \sim 0.562$ and $k_{conv} \sim 0.13$, given that the angle of convergence of the pipe is smoother, 86.5°, instead of 80° for which $k_{conv} \sim 0.2$, as reported in Çengel and Cimbala (2010). Using as average velocity the one in the smallest cross section of the Minoan pipe $u_m \equiv 1.4$ m/s, Eq. (10) gives $\Delta P/L = 941$ Pa/m, a much higher value than the value for a straight pipe, and by ~30% higher than the numerically computed value of 725 Pa/m, given in Table 5. Hence these approximate expressions provide acceptable estimates of the energy loss in these geometries. On the other hand, they cannot help us visualize the flow and thus understand the underlying physics and the operation of the pipes. Having shown numerically that even perfectly smooth pipes are not energetically efficient and with approximate analysis that non-smooth pipes are even less efficient, it is considered redundant to numerically compute the flow in non-smooth pipes.

The so-called wall variables are ubiquitous in the turbulence literature. In Fig. 8 we give the variation of the dimensionless distance from the wall, y^+, measured in terms of the characteristic length scale

$\mu / \dfrac{\sqrt{\rho \tau_0}}{\mu}$, where τ_0 is the shear stress at the wall:

$$y^+ = y \frac{\sqrt{\rho \tau_0}}{\mu}$$

(11)

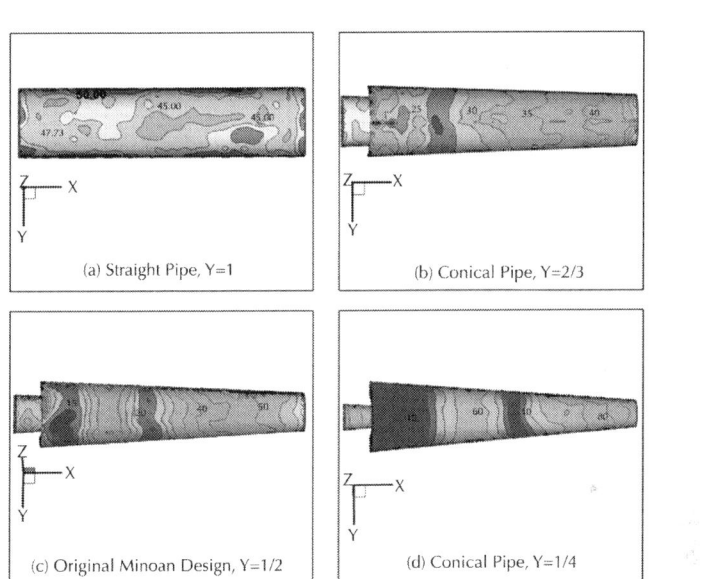

(a) Straight Pipe, Y=1

(b) Conical Pipe, Y=2/3

(c) Original Minoan Design, Y=1/2

(d) Conical Pipe, Y=1/4

Figure 8: Effect of pipe geometry on the contours of y^+ on the walls for four different designs (for Q = 8 l/s): a) Straight pipe, Y = 1, b) Conical pipe with Y = 2/3, c) Original Minoan design, Y = 1/2, d) Conical pipe with Y = 1/4.

The full range of the wall coordinate is $0 \le y^+ \le y_c^+$, where y_c^+ corresponds to the location of the pipe center and is given by $y_c^+ = \text{Re}\sqrt{f/32}$. In Fig. 8 we observe that for a straight pipe $y^+ \approx 45-47$ throughout the wall. As Y decreases, y^+ varies monotonically along the wall, increasing as the cross section decreases. Additionally, near the abrupt expansion it assumes its smaller value. This value decreases as Y decreases, because the latter decrease leads to stronger recirculation resulting in a thinner laminar sublayer near the wall. As the pipe exit is approached, y_c^+ increases and more rapidly when the pipe radii decrease more, because, although the viscous sublayer gets thinner as the velocity increases, it occupies a larger fraction of the tube cross section when Y = 1/4.

Effect of Flow Rate on the Flow Fields

We have also examined the effect of the flow-rate on the flow field. To this end, we have performed additional simulations with higher flow rates Q ≈ 11 l/s and Q ≈ 14 l/s. The values of 8 l/s and 14 l/s correspond to the expected minimum and maximum flow rates for a Minoan water distribution system (Dialynas et al., 2006). Apparently, the recirculation downstream the pipe entrance gets more extended and intense as the flow rate increases (see Fig. 9). As expected, the magnitude of the mean velocity $\underline{U}$, as well as the magnitude of the velocity fluctuations $|\underline{u}'|$ in the form of a Reynolds number, and the strain rate reach substantially higher values for the maximum value of the flow rate. Interestingly, although the relative turbulence scale (turbulence intensity) is not significantly affected by the increase of the flow rate, the absolute turbulence scale, measured via the turbulence Re, clearly increases. The turbulence Reynolds number is defined as follows:

$$Re_I = \frac{|\underline{u}'|\rho D}{\mu}$$

(12)

where $|\underline{u}'|$ is the magnitude of the fluctuation velocity given in Eq. (6). This different behavior arises because turbulence intensity is a relative measure with respect to mean velocity, whereas the turbulence Reynolds number is an absolute measure of the flow intensity.

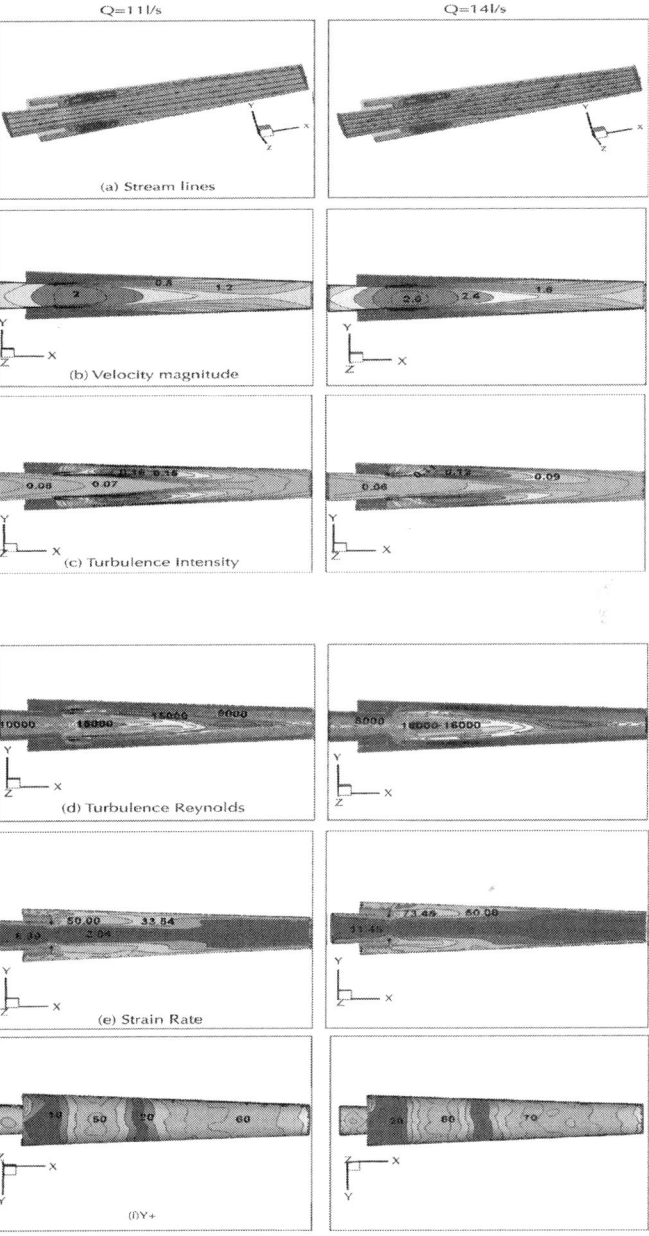

Figure 9: Effect of the flow rate on the flow characteristics for flow rate values of 8 l/s (left column) and 14 l/s (right column) (for Y = 1/2 and gravity normal to the direction of the flow) a) Streamlines, b) Velocity magnitude in [m/s] c) Turbulence intensity, d) Turbulence Reynolds, e) Strain rate in [s⁻¹].

As far as the effect of the flow rate to the pressure drop and the drag force is concerned, we can discern a rise of about ~200% in these variables, when the flow rate increases from 8 l/s to 14 l/s (Table 6). Observing Eq. (8) it should have been expected that ΔP/L will vary with Q^2. Indeed, the examined increase in Q from 8 l/s to 14 l/s yields the value $(14/8)^2 = 3.1$, while the computed ratio in corresponding pressure drops fromTable 6 is 2170/725 = 3.0, in accordance with Eq. (8).

Table 6: Effect of flow rate on pressure drop and drag coefficient for Y = 1/2

Q (l/s)	P/L (Pa/m)	FD (N)
8 l/s	−724.982	3.406
11 l/s	−1347.98	6.429
14 l/s	−2169.58	10.194

Effect of Pipe Orientation

As far as the orientation of the pipe segment is concerned, with respect to the gravity field, we choose four different orientations expressed by the angle Φ, which is formed between the gravity field and the axis of the pipe in the flow direction. For the value of $\Phi = 0°$ the flow direction is along with the orientation of the gravity field, for $\Phi = 180°$ the flow direction is opposite to it and for $\Phi = 90°$ the flow is normal to the gravity, which is the arrangement we discussed so far. As last case we examine the flow direction forming a $\Phi = 30°$ angle with the gravity field.

Despite the fact that the pipe orientation was changed, the contours of the velocity magnitude remain unchanged from those reported previously, because the same flow rate at the entrance is imposed in all orientations. We also observe that the drag force itself remains nearly unchanged as the orientation of the pipe changes, since the flow rate and consequently the velocity inside the pipe remains unchanged in these cases. However, as seen in Table 7, the pressure drop along the pipe segment is greatly affected. When $\Phi = 0$ the flow is driven by gravity and to keep the flow rate at $Q \approx 8$ l/s an opposing the flow pressure difference is needed. Qualitatively the same is the situation with = 30°. Only when the pipe axis is vertical to the gravity direction the flow requires a pressure drop to be generated, while P/L increases significantly when it drives the flow in the direction opposite to gravity.

Table 7: Effect of pipe inclination with respect to gravity for Q = 8 l/s and Y = 1/2

Φ	ΔP/L (Pa/m)	FD (N)
0°	1498.174	3.412
30°	1125.557	3.420
90°	−724.982	3.406
180°	−2897.099	3.401

CONCLUSIONS

The primary objective of this work was to characterize the operation and present the physical phenomena that occur inside the flow of conical Minoan Terracotta pipes and, hence, determine the reason of their usage and their advantages or deficiencies.

The various historical data and the information that were collected indicate that the Minoan Terracotta pipes were without question one of the stepping stones in water distribution technology of ancient civilizations. They were used to transport water in complex distribution systems in ancient Knossos and other glorious Minoan cities of the Bronze Age and provide water in fortified settlements. These pipes raise questions even nowadays concerning the knowledge of the ancient civilizations about hydraulics. Moreover, mystery covers the disappearance of these pipe designs in later times, leading to disputes between modern researchers. What is the main reason that the Minoans initially chose the conical design with the specific type of joints for their pipelines? Had there been a technical disadvantage that led to the disposal of the Minoan Terracotta pipes, or was the knowledge of their manufacturing simply lost with the decadence and the destruction of the Minoan civilization?

In order to address some of the above issues or to exclude some of the prior hypotheses on this matter, we simulated the flow inside the conical pipe design of the ancient Minoans and compared its behavior with different geometries under the assumption of periodic flow. Because of the high flow rate, the flow is turbulent and the Reynolds-averaged Navier–Stokes equations along with the eddy viscosity k–ε model are adopted for the physical modeling of the process. The calculations show that the main disadvantage of the conical pipes was the large amount of head loss (Table 5) that is specifically due to the geometry of the joints between the segments. The turbulence intensity in Minoan pipes is twice that in cylindrical tubes due to the destabilizing effect of boundary layers that get detached downstream of the entrance section of each tube segment. This hydraulic disadvantage might have led to a substantial loss of water volume when the sealing is not tight enough and failure of this geometry for large water distribution systems. Moreover, the contours of the stream function show that wide recirculation (Fig. 4) develops in the annular part of the tube around the edge of the introductory section. Apparently this could lead to increased entrapment of sediments in this region due to the formation of closed flow loops. On the other hand, the high flow intensity could prevent deposition of these sediments on the pipe wall. Our results concerning the orientation of the pipe segments show that, as expected, the energy loss would be acceptable if the pipe was placed in a way that the water flow was directed "downhill", however the high losses would decelerate the "uphill" flow in the pipes.

Despite their dysfunction, the broad usage of the conical pipes throughout the Minoan kingdom implies that their initial usage was based on a logical and viable acclaim. Our analysis leads us to hypothesize that the main advantage of this geometry is the simplicity of the construction of the pipes and their assembling to form networks, as there was no need for separate fittings and angled pipe parts, in order to create curves in the distribution system. Let us not forget that taking under consideration the materials (clay) and the production capabilities that were available at that time, a single production line of pipe segments, instead of multiple designs (additional fittings and angular parts) was a more viable option technologically and economically. Other reasons for choosing this design, which are more related to the findings of this study could be the following: (1) the strong recirculation in the pipe entrance could prevent from deposits accumulating in the pipe wall and the increased turbulence level could carry away small particles more efficiently and (2) the small area behind the recirculation, where the velocity is much smaller could be used to open a small hole and, thus, vent a horizontal pipe to the atmosphere from its upper surface without excessive leakage of water or clean and maintain it. Venting the pipeline in such a way is known to prevent water bursts from propagating downstream. Such holes have been reported in several pipelines in ancient Athens (Angelakis et al., 2005) or in later Roman times in Ephesos (Ortloff and Crouch, 2001) and in Aspendos (Ortloff and Kassinos, 2003).

ACKNOWLEDGMENTS

The authors would like to thank both reviewers for their constructive comments, which helped them improve the presentation of their work.

REFERENCES

1.   Angelakis, A.N., Spyridakis, D.S., 2010. A brief history of water supply and wastewater management in ancient Greece. Water Science and Technology: Water Supply 10 (4), 618e628.

2.   Angelakis, A.N., Koutsoyiannis, D., Tchobanoglou, G., 2005. Urban wastewater and stormwater technologies in ancient Greece. Water Research 39 (1), 210e220.

3.   Angelakis, A.N., Savvakis, Y.M., Charalampakis, G., 2007. Aqueducts during the Minoan Era. Water Science and Technology: Water Supply 7 (1), 95e101.

4.   Ball, P., 2003. Ancient aqueduct at Aspendos was a hydraulic masterpiece. Nature. http://dx.doi.org/10.1038/news030407-3 (published online).

5.   Çengel, Cimbala, 2010. Fluid Mechanics. Mc Graw Hill.

6.   Deen, W.M., 1998. Analysis of Transport Phenomena. Oxford Univ. Press.

7.   Dialynas, E., Lyrintzis, A., Angelakis, A.N., 2006. Historical development of water supply in Iraklio City, Greece. In: Proceedings of 1st IWA International Symposium on Water

and Wastewater Technologies in Ancient Civilizations, 28e30 October 2006, Iraklio, Greece, pp. 671e676.

8. Dimakopoulos, Y., Tsamopoulos, J., 2006. Gas assisted injection molding with fluids partially occupying straight and complex tubes. Polymer Engineering & Science 46 (1), 47e68.

9. Eker, _ I., Grimble, M.J., Kara, T., 2002. Operation and simulation of city of Gaziantep water supply in Turkey. Renewable Energy 28 (6), 901e916.

10. FluentTM. ANSYS, Inc., 2012. http://www.ansys.com.

11. Haaland, S.E., 1983. Simple and explicit formulas for the friction factor in turbulent pipe flow. Journal of Fluids Engineering 105 (1), 89e90.

12. Haut, B., Viviers, D., 2007. Analysis of the water supply system of the city of Apamea, using Computational Fluid Dynamics. Hydraulic system in the north-eastern area of the city, in the Byzantine period. Journal of Archaeological Science 34 (3), 415e427.

13. Kouris, Ch., Dimakopoulos, J., Georgiou, G., Tsamopoulos, J., 2002. Comparison of spectral and finite element methods applied to the study of interfacial instabilities of the core-annular flow in an undulating tube. International Journal Numerical Methods in Fluids 39 (1), 41e73.

14. Mays, L.W., 2008. A very brief history of hydraulic technology during antiquity. Environmental Fluid Mechanics 8, 471e484.

15. Mays, L.W., 2010. A brief history of water technology during antiquity: before the Roman. In: Mays, L.W. (Ed.), Ancient Water Technologies. Springer, pp. 1e28. ISBN:987-90-481-8631-0.

16. Nicolic, M., 2008. Cross Disciplinary Investigation of Ancient Long-distance Water Pipelines. PhD thesis, University of Victoria.

17. Ortloff, C.R., Crouch, D.P., 2001. The urban water supply and distribution system of the Ionian city of Ephesos in the Roman imperial period. Journal of Archaeological Science 28, 843e860.

18. Ortloff, C.R., Kassinos, A., 2003. Computational fluid dynamics investigation of the hydraulic behavior of the Roman inverted siphon system in Aspendos, Turkey. Journal of Archaeological Science 30, 417e428.

19. Papaioannou, J., Karapetsas, G., Dimakopoulos, Y., Tsamopoulos, J., 2009. Injection of a viscoplastic material inside a tube or between two parallel disks: conditions for wall detachment of the advancing front. Journal of Rheology 53 (5), 1155e1191.

20. Pavlidis, M., Dimakopoulos, Y., Tsamopoulos, J., 2010. Steady viscoelastic film flow over 2D topography: I. The effect of viscoelastic properties under creeping flow. Journal of Non-Newtonian Fluid Mechanics 165 (11e12), 576e 591.

21. Poroseva, S.V., 2001. Modeling the "rapid" part of the velocity-pressure gradient correlation in inhomogeneous turbulence. In: Annual Research Briefs. Center for Turbulence Research, NASA Ames/Stanford Univ., pp. 367e374.

22. Pujol, T., Montoro, L., 2010. High hydraulic performance in horizontal waterwheels. Renewable Energy 35 (11), 2543e2551.

23. Webster, L., Hughes, R., 2010. The mystery of Minoan tapered pipes. Department of Civil and Environmental Engineering, The University of Melbourne, Victoria, Australia. HydroLink 5 (2), 27e29.

24. Yih, C.-S., 1977. Fluid Mechanics. West River Press, Ann Arbor Michigan, USA.

Numerical Simulation on Dense Phase Pneumatic Conveying of Pulverized Coal in Horizontal Pipe at High Pressure

Wenhao Pu[a, b], Changsui Zhao[b], Yuanquan Xiong[b], Cai Liang[b], Xiaoping Chen[b], Peng Lu[b], and Chunlei Fan[b]

[a]College of Energy and Power Engineering, Nanjing University of Aeronautics and Astronautics, Nanjing 210096, China
[b]School of Energy and Environment, Southeast University, Nanjing 210096, China

ABSTRACT

A kinetic–frictional model, which treats the kinetic and frictional stresses in an additive manner, was incorporated into the two fluid model based on the kinetic theory of granular flow to simulate three dimensional flow behaviors of dense phase pneumatic conveying of pulverized coal in horizontal pipe. The kinetic stress was modeled by the kinetic theory of granular flow, while the friction stress is from the combination of the normal frictional stress model proposed by Johnson and Jackson [1987. Frictional–collisional constitutive relations for granular materials, with application to plane shearing. Journal of Fluid Mechanics 176, 67–93] and the modeled frictional shear viscosity model proposed by Syamlal et al. [1993. MFIX documentation and theory guide, DOE/METC94/1004, NTIS/DE94000087. Electronically available from http://www.mfix.org], which was modified to fit experimental data. For the solid concentration and gas phase Reynolds number was high, the gas phase and particle phase were all treated as turbulent flow. The experiment was carried out to validate the prediction results by three kinds of measurement methods. The predicted pressure gradients were in good agreement with experimental data. The predicted solid concentration distribution at cross section agreed well with electrical capacitance tomography (ECT) image, and the effects of superficial velocity on solid concentration distribution were discussed. The formation and motion process of slug flow was demonstrated, which is similar to the visualization photographs by high speed video camera.

INTRODUCTION

High pressure and large-scale entrained bed gasification technology is one of the promising coal gasification technologies for IGCC, indirect coal liquefaction, chemical fertilizer, hydrogen production and fuel cell, etc, while dense phase pneumatic conveying of pulverized coal at high pressure is one of the most important techniques in coal gasification process. Due to the inherent complexity of dense phase pneumatic conveying in horizontal pipe, numerical simulation can provide valuable insights into particle flow process, which are difficult to be obtained from experiments.

In the framework of Eulerian approach, the solid phase was commonly modeled with the kinetic theory of granular flow, which was adopted by some authors (Lun et al., 1984; Sinclair and Jackson, 1989; Ocone et al., 1993; Benyahia et al., 2000) to simulate various gas solid flows. In the past few decades, the dilute flow had been simulated with success based on the classic kinetic theory. Some researchers (Van Wachem et al., 2001; Van Wachem and Almstedt, 2003; Curtis and van Wachem, 2004; Gidaspow et al., 2004) reviewed the kinetic theory of granular flow. However, some shortcomings have been pointed out when the dense flow was predicted by employing the classic kinetic theory. Several researchers (Sundaresan, 2000;McKeen and Pugsley, 2003; Makkawi et al., 2006) attributed these to the fact that the classic kinetic theory of granular flow only accounted for the particle kinetic stress, with no account of other possible contact stress such as inter-particle cohesion or frictional stress. In regions with high particle volume fractions, the dominant stress generation mechanism is more likely to be due to long-term and multi-particle contacts. In recent years, the modeling of dense flow using kinetic–frictional stress models (Syamlal et al., 1993; Huilin et al., 2004; Patil et al., 2005a and Patil et al., 2005b) has flourished. The frictional stress models mostly originated from geological research groups and followed a rigid-plastic rheology assumption (Schaeffer, 1987; Johnson and Jackson, 1987; Tardos, 1997). Following Savage (1998), the frictional stress models were simply added to the kinetic stresses as described by kinetic theory. Although there was no physical basis for this additive assumption, dense-phase flow behavior over a range of solid volume fractions seemed to be accurately captured by employing this assumption. Huilin et al. (2004) found that if the frictional stress is omitted, the porosity in the annular zone of a spouted fluidized bed is over-predicted.Makkawi et al. (2006) reached the same conclusion when comparing the experimental data of bubbling bed with the prediction of the MFIX code (Syamlal et al., 1993), where the frictional stress was not included.Srivastava (2001) and Srivastava and Sundaresan (2003) developed a kinetic–frictional model based on the work of Savage (1998), and found that this model incorrectly predicted the behavior of stagnant regions and over-predicts hopper discharge rate.

Most of these investigations discussed above shared the same characteristic that the gravity was parallel to the flow direction, involving bubbling beds, circulating fluidized beds, spouted bed and hoppers. In horizontal pneumatic conveying, the flow features were very complicated because of the action of gravity perpendicular to the flow direction (Zhu et al., 2004). Particles were suspended in the gas in the upper portion of the pipe. At the bottom the particles deposited forming a settled layer. In the former, the kinetic stress played an important role, while in the latter with high solid concentration the dominant stress generation mechanism was more likely to be due to long-term and multi-particle contacts. In long-term contacts much more energy will be dissipated which results in a self-enhancing mechanism for the formation of extremely dense regions, since these particles have hardly any energy left to escape these regions.

There were some researches on simulation of dense phase horizontal pneumatic conveying. Tsuji et al. (1992) applied discrete element method (DEM) for plug flow simulation in a horizontal pipe. Due to the computation limitations, only a very short pipe (0.6 m), large particles (10 mm) and a small amount of particles (150) in the simulation flow were considered. Recently, the DEM was evolved to treat large-scale powder systems (Sakai and Koshizuka, 2009; Tsuji et al., 2008). To reduce the calculation cost, Sakai and Koshizuka (2009) presented a coarse grain model for large-scale DEM simulations, which was applied to a three-dimensional plug flow in a horizontal pipeline. Tsuji et al. (2008) applied discrete element method to investigated the "scale effect" of the gas-fluidized bed size and obtained the three-dimensional bubble structures, in which a maximum of 4.5 million particles were tracked by using 16 CPUs of a commodity PC cluster computer. Zhu et al., 2008 and Zhu et al., 2007 reviewed the major theoretical developments and the applications of DEM. It made a great progress, but the calculation cost was still much high for the real gas solid systems. For the dense phase pneumatic conveying of pulverized coal at high pressure, the particle size was about tens of micrometer and the number of the particles was far more than millions, so this method was not suitable for our simulation.

Levy (2000) applied the two-fluid theory for modeling the slug formation and breakage processes. The results showed the slugs increased in length along the pipeline and further downstream they broke up in a series of short slugs being close together. The solid phase stress was modeled using an empirical equation in his simulations. Tardos et al. (2003) proposed a new expression for the average solid shear stress that smoothly merged the slow-intermediate regime with the rapid flow regime. Makkawi and Ocone (2006)proposed a one-dimensional fully developed flow model, which incorporated expressions presented byTardos et al. (2003) into kinetic theory, and predicted the gas solid flow in a horizontal duct successfully. But the model was difficult to extend to three dimensional flow simulations because of the comparison between kinetic stress and frictional stress. However, all these simulations were validated by qualitative comparison and lack of the experimental data of solid volume fraction distribution.

Konrad (1986) reviewed some definitions about the dense phase. He concluded that there was no universally accepted definition of dense phase conveying, and pointed out that the solids loading ratio or the solid concentration could not be the criteria. When the gas phase was insufficient to support all the particles in suspension, it was regarded as dense phase conveying. That is to say, stratified flow, slug flow and plug flow were all regarded as dense phase flow. Because of the requirement of the gasifier, the solid feeding was truly continuous in our experiments. It was different from the plug flow, which was separated by the gas phase. In experiments, the solids loading ratio ranged from 7.54 to 15.26 and the solid concentration was between 0.16 and 0.33. The solids loading ratio was not very large. Compared with the pneumatic conveying at atmospheric pressure, the transport pressure was much larger and ranged from 2.6 to 3.3 MPa. For example, when the transport pressure was 2.6 MPa, the gas density and the gas mass flow rate were 26 times larger than those at atmospheric pressure. So the solids loading ratio became less. But on the other hand, the ratio of solid mass flow rate to gas volume flow rate was between 239 and 506 kg/m^3 and the solid mass flux was over 2100 kg/(m^{2}s). These data showed the advantages of our experimental setup. In our simulations, we focused on that how to describe the deposited particles and the suspended particles simultaneously.

There were almost no detailed published data on solid concentration distribution in horizontal pipe, due to not only their spatial and temporal complexities but also practical difficulties of the measurement. Differential pressure transducers were usually used to measure the pressure drop; however, the flow information was over-simplified. Traditionally, the flow instabilities presented in the dense flow have been investigated by the use of photography. This enabled extracting the shape of the flow structures, but it must be remembered that particles can remain in each other's shadows and the data interpretation is difficult (Jaworski and Dyakowski, 2002). Moreover, no information was available on the internal structure of the flow such as solid volume fraction distribution. It is worth noting that the Electrical Capacitance Tomography (ECT) becomes a promising technique to overcome these weaknesses in recent years. In order to validate the simulation results, three kinds of measurement methods were employed to get more useful information in experiments.

Only a few researches (Geldart and Ling, 1990; Chen et al., 2007; Liang et al., 2007; Pu et al., 2008) on dense phase pneumatic conveying at high pressure have been reported in the literatures so far. Among them, little work on simulation of slug flow in horizontal pipe was reported. Here a kinetic–frictional model, which was revised according to the pressure gradients from experiments, was attempted to predict the slug flow for dense phase pneumatic conveying at high pressure. In dense phase pneumatic conveying in horizontal pipe at high pressure, the average solid concentration was up to 30% at the pressure of 3.3 MPa in a confined space with high transport velocity. Thus it is necessary to consider both kinetic stress and frictional stress. At the same time, the particle phase and the gas phase were all treated as turbulent flow due to high gas phase Reynolds and solid concentration. Further details can be founded in literatures (Zheng et al., 2001; Pu et al., 2008).

A kinetic–frictional model, which considered the kinetic and frictional stresses of the particle phase and the turbulence interaction between gas phase and particle phase, was incorporated into the two fluid model based on the kinetic theory of granular flow to simulate three dimensional flow behaviors of dense phase pneumatic conveying of pulverized coal in horizontal pipe. The kinetic stress was based on the kinetic theory of granular flow. For frictional stress, a normal frictional stress model proposed by Johnson and Jackson (1987) and the modified frictional shear viscosity model proposed by Syamlal et al. (1993) were used, which was modified to fit experimental data. The experiments were carried out to validate the prediction results. Firstly, the predicted pressure gradients were compared with the experimental data. Secondly, the solid concentration distribution at cross section was obtained and compared with the measurements by ECT. Then the effects of superficial velocity on solid concentration distribution were discussed. Finally, the formation and motion process of slug flow was simulated and compared with the photographs by high speed camera.

EXPERIMENTAL SETUP

The experiment apparatus used in this study is illustrated in Fig. 1. The pulverized coal with average particle diameter of 37 μm and real density of 1350 kg/m³ was conveyed between two pressurized hoppers (3) via a 53 m long and 10 mm inside diameter pipe. As conveying gas, the nitrogen was introduced into the system via a header (7) connected with sixteen nitrogen cylinders (8) and then divided into three parts: pressurizing gas (4), fluidizing gas (5) and supplemental gas (6). The pressurizing gas (4) was to maintain the stabilization of pressure in the sending hopper. The fluidizing gas (5) fluidized the pulverized coal at the bottom of the sending hopper, and then drove the pulverized coal into the conveying pipe. The supplemental gas (6) was used to adjust the nitrogen flow rate in a wide range and prevent pipe from blockage to guarantee continuous transport of pulverized coal. The pressure in the receiving hopper was controlled by motor-driven valve (1). Each of the three nitrogen flow rates was measured and controlled by a metal rotor flow meter. The dosing hopper (3) was suspended on three load cells (2), by which the pulverized coal mass flow rate was measured. The EJA 430A differential pressure transducers (11) were used to measure the pressure drops of sections in horizontal pipe, vertical pipe and two bends. A computer data acquisition system was employed to record the data of flow rates and pressure drops.

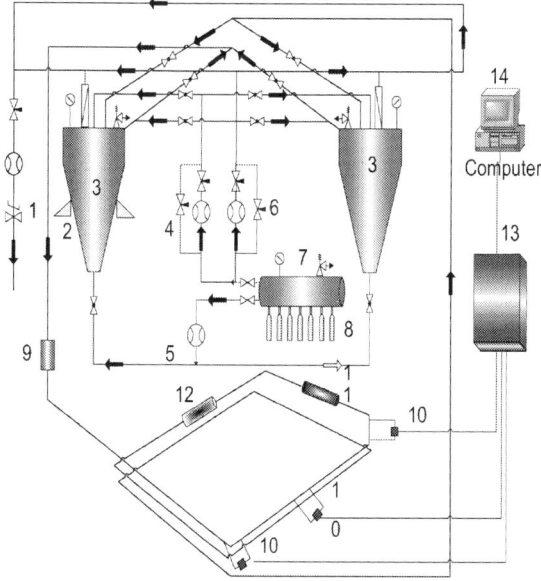

Figure 1: Schematic diagram of the dense phase pneumatic conveying system of pulverizedcoal (1) motor-driven valve (2) weigh cells (3) hoppers (4)

pressurizing gas (5) fluidizing gas (6) supplemental gas (7) header (8) nitrogen cylinder (9) adding water apparatus (10) differential pressure transducer (11) ECT sensor (12) visualization section (13) data acquisition system (14) computer.

The ECT sensor consists of eight equally spaced sensing electrodes of 10 mm long, is shown schematically in Fig. 2 (a). The sensor pipe is made of quartz glass with 20 mm outer diameter. The ECT sensor can provide dynamic information on the spatial distribution of the composite permittivity of the gas solid phase at 40 frames per second. From the permittivity data, the pipe cross-sectional relative distribution of the two phases can be obtained. Further details on the ECT sensor design can be found inYang (2008). Sensor calibration is performed by taking the capacitance readings for two situations: an empty pipe and the pipe filled with pulverized coal. The high speed video camera FASTCAM-NET-MAX3 is employed to photograph the flow regimes through the transparent pipe during the experiments. Two groups of 2000 W floodlights were used to enhance the photo definition when photographing. The visualization section made of quartz glass with 10 mm inside diameter is shown in Fig. 2(b).

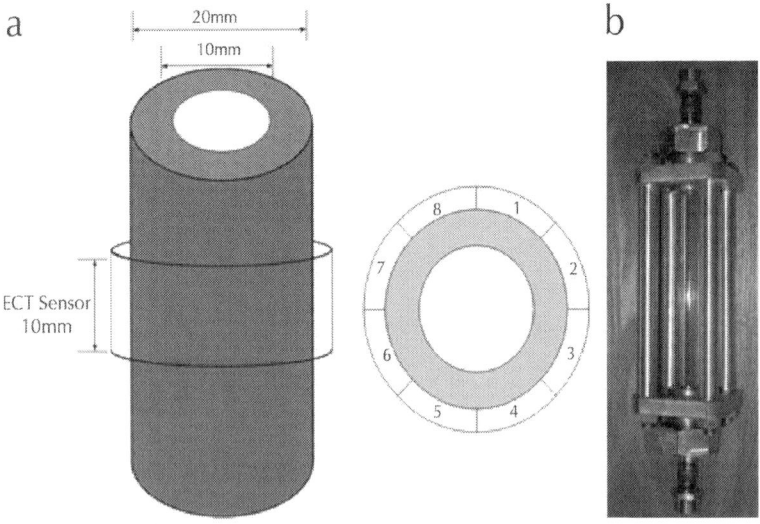

Figure 2: (a) Schematic of the ECT sensor and (b) visualization section.

HYDRODYNAMIC MODEL

The model adopted is based on the fundamental concept of interpenetrating continua for multiphase mixtures. It is assumed that different phases can be present at the same time in the same computational volume. The fundamental equations of mass, momentum, and energy conservation are solved for each considered phase. The emerging kinetic theory of granular flow provides a physical motivation for such an approach. Appropriate constitutive equations have to be specified in order to describe the physical and rheological properties of each phase and to close the conservation equations.

Gas phase

The continuity, momentum equations for gas phase and the constitutive relations:

Gas phase continuity equation

$$\frac{\partial}{\partial t}(\alpha_g \rho_g) + \nabla \cdot (\alpha_g \rho_g \mathbf{v}_g) = 0$$

(1)

Gas phase momentum equation

$$\frac{\partial}{\partial t}(\alpha_g \rho_g \mathbf{v}_g) + \nabla \cdot (\alpha_g \rho_g \mathbf{v}_g \mathbf{v}_g) = -\alpha_g \nabla p_g + \nabla \cdot \tau_g - \beta(\mathbf{v}_g - \mathbf{v}_s) + \alpha_g \rho_g \mathbf{g}$$

(2)

where, $_g$ is the gas phase volume fraction, $_g$ is the gas phase density, $_g$ is the gas phase velocity, g is the specific gravity force. $_g$ is the gas phase shear stress and is related to the gradients of velocity components by

$$\tau_g = \alpha_g \mu_{g,eff}[\nabla \mathbf{v}_g + (\nabla \mathbf{v}_g)^T] + \alpha_g(\lambda_g - \tfrac{2}{3}\mu_{g,eff})(\nabla \cdot \mathbf{v}_g)\mathbf{I}$$

(3)

where, $\mu_{g,eff}$ is the effective gas phase viscosity and $\mathbf{I}$ is the unit tensor.

In dense gas–solid systems, the dominant contributor to the interaction term is the drag force. Here is the drag coefficient of gas–solid flow, defined as

$$\beta = \begin{cases} \dfrac{3}{4} C_D \dfrac{\alpha_s \alpha_g \rho_g |\mathbf{v}_g - \mathbf{v}_s|}{d_s} \alpha_g^{-2.65}, & \alpha_g > 0.8 \\[3mm] 150 \dfrac{\alpha_s^2 \mu_g}{\alpha_g d_s^2} + 1.75 \dfrac{\rho_g \alpha_s}{d_s} |\mathbf{v}_g - \mathbf{v}_s|, & \alpha_g \leq 0.8 \end{cases}$$

(4)

$$C_D = \begin{cases} \dfrac{24}{Re_s}[1 + 0.15(Re_s)^{0.687}], & Re_s < 1000 \\[3mm] 0.44, & Re_s \geq 1000 \end{cases}$$

(5)

and the relative Reynolds number, Re_s is defined by

$$Re_s = \frac{\alpha_g \rho_g d_s |\mathbf{v}_s - \mathbf{v}_g|}{\mu_g}$$

(6)

Particle Phase Kinetic Theory

The continuity, momentum equations for particle phase and the constitutive relations are expressed as

Particle phase continuity equation

$$\frac{\partial}{\partial t}(\alpha_s \rho_s) + \nabla \cdot (\alpha_s \rho_s \mathbf{v}_s) = 0$$

(7)

Particle phase momentum equation

$$\frac{\partial}{\partial t}(\alpha_s\rho_s\mathbf{v}_s)+\nabla\cdot(\alpha_s\rho_s\mathbf{v}_s\mathbf{v}_s)=-\alpha_s\nabla p_g+\nabla\cdot\tau_s+\beta(\mathbf{v}_g-\mathbf{v}_s)+\alpha_s\rho_s\mathbf{g}$$

(8)

where, $_s$ is the particle phase stress. Following Savage (1998), it is assumed that the particle phase stress tensor $_s$ is simply the sum of the kinetic stress tensor τ_s^k and the frictional stress tensor τ_s^k, each contribution evaluated as if it acted alone:

$$\tau_s = \tau_s^k + \tau_s^f$$

(9)

Although this approach has no strong physical justifications, nevertheless it captures the two extreme limits of granular flow the rapid shear flow regime where kinetic contributions dominate and the quasi-static flow regime where friction dominates. It produced good results for dense flow (Srivastava and Sundaresan, 2003).

Kinetic Stress Model

The kinetic stress tensor is now commonly modeled by the kinetic theory of granular flow (Gidaspow, 1994). Solid viscosity and pressure are derived by considering the random fluctuation of solid velocity and its variations due to particle–particle collisions. Such a random kinetic energy, or granular temperature, can be predicted by solving, in addition to the mass and momentum equations, a fluctuating kinetic energy equation for the particles. The solid viscosity and pressure can then be computed as a function of granular temperature at any time and position. Particles are considered smooth, spherical, inelastic, and undergoing binary collisions. Closure relations for τ_s^k and k_s used here are presented in Table 1. The solid pressure represents the normal forces due to particle–particle interactions. The first part of the solid pressure represents the kinetic contribution, and the second part represents the collisional contribution. The kinetic part physically represents the momentum transferred through the system by particles moving across imaginary shear layers in the flow; the collisional part denotes the momentum transferred by direct collisions. The solid bulk viscosity describes the resistance of the particle suspension against compression. Where e_{ss} is the restitution coefficient of particle–particle collision and $g_{0,ss}$ is the radial distribution function expressing the statistics of the spatial arrangement of particles. Particles are in constant contact at the maximum solid volume fraction, the radial distribution function at contact tends to infinity. It approaches the correct limit of one as the solid volume fraction approaches zero. Van Wachem et al. (2001) compared four forms of the radial distribution function. Here the expression of Sinclair and Jackson (1987) is adopted.

Table 1: Kinetic theory of granular flow closures

$$\tau_s^k = (-p_s^k + \alpha_s \lambda_s^k (\nabla \cdot \boldsymbol{v}_s)) \boldsymbol{I} + 2\alpha_s \mu_s^k \boldsymbol{S}_s$$

$$\boldsymbol{S}_s = \tfrac{1}{2}(\nabla \boldsymbol{v}_s + (\nabla \boldsymbol{v}_s)^T) - \tfrac{1}{3}(\nabla \cdot \boldsymbol{v}_s)\boldsymbol{I}$$

$$p_s^k = \alpha_s \rho_s \Theta_s + 2\rho_s(1 + e_{ss})\alpha_s^2 g_{0,ss}\Theta_s$$

$$\mu_s^k = \tfrac{1}{2}\alpha_s \rho_s d_s g_{0,ss}(1 + e_{ss})\left(\frac{\Theta_s}{\pi}\right)^{1/2} + \frac{10\rho_s d_s \sqrt{\Theta_s \pi}}{96(1 + e_{ss})g_{0,ss}}\left[1 + \tfrac{4}{5}(1 + e_{ss})\alpha_s g_{0,ss}\right]^2$$

$$\lambda_s^k = \tfrac{4}{3}\alpha_s \rho_s d_s g_{0,ss}(1 + e_{ss})\left(\frac{\Theta_s}{\pi}\right)^{1/2}$$

$$g_{0,ss} = [1 - (\alpha_s / \alpha_{s,max})^{1/3}]^{-1}$$

$$k_{\Theta_s} = \frac{150\rho_s d_s \sqrt{\Theta_s \pi}}{384(1 + e_{ss})g_{0,ss}}\left[1 + \tfrac{6}{5}\alpha_s g_{0,ss}(1 + e_{ss})\right]^2 + 2\alpha_s^2 \rho_s d_s g_{0,ss}(1 + e_{ss})\sqrt{\frac{\Theta_s}{\pi}}$$

$$\gamma_{\Theta_s} = 3(1 - e_{ss}^2)g_{0,ss}\rho_s \alpha_s^2 \Theta_s\left(\frac{4}{d_s}\sqrt{\frac{\Theta_s}{\pi}} - \nabla \cdot \boldsymbol{v}_s\right)$$

$$\varphi_{sg} = -3\beta\Theta_s$$

Particle phase pseudo-temperature equation of the conservation of particles fluctuating energy is given by

$$\frac{3}{2}\left[\frac{\partial}{\partial t}(\rho_s \alpha_s \Theta_s) + \nabla \cdot (\rho_s \alpha_s \boldsymbol{v}_s \Theta_s)\right] = \tau_s^k : \nabla \boldsymbol{v}_s + \nabla \cdot (k_{\Theta_s}\nabla \Theta_s) - \gamma_{\Theta_s} + \phi_{sg} \tag{10}$$

where Θ_s is the particle pseudo-temperature, k_{Θ_s} is the transport coefficient of pseudo-thermal energy. The first term on the right-hand side of this equation represents the production rates of pseudo-thermal energy by shear. The second one represents the diffusive transport of pseudo-thermal energy. The third term $\gamma_{\Theta s}$ in the equation represents dissipation of pseudo-thermal energy through inelastic collisions, whereas the fourth term Φ_{sg} denotes the exchange of fluctuating energy between gas and particles.

Frictional Stress Model

At high solid volume fraction, individual particle interacts with multiple neighbors through sustained contact. The resulting frictional stress must be accounted for in the description of the solid phase stress. The frictional stress is expressed in Table 2. For detailed information, please refer to Syamlal et al. (1993)and Srivastava and Sundaresan (2003).

Table 2: Frictional stress model

$$\tau_s^f = -p_f \mathbf{I} + 2\alpha_s \mu_f \mathbf{S}_s$$

$$\mu_f = \frac{p_f \sin\phi}{\sqrt{2}\alpha_s \sqrt{\mathbf{S}_s : \mathbf{S}_s + \Theta_s/d_s^2}} \left\{ n - (n-1)\left(\frac{p_f}{p_c}\right)^{1/n-1} \right\}$$

$$n = \begin{cases} \dfrac{\sqrt{3}}{2\sin\phi} & \nabla \cdot \mathbf{v}_s > 0 \\ 1.03\nabla \cdot \mathbf{v}_s < 0 \end{cases}$$

$$p_f = \left[1 - \frac{\nabla \cdot \mathbf{v}_s}{\sqrt{2}n\sin\phi\sqrt{\mathbf{S}_s : \mathbf{S}_s + \Theta_s/d_s^2}} \right]^{n-1} p_c$$

$$p_c = \begin{cases} 10^{24}(\alpha_{s,max} - \alpha_s)^{10} & \alpha_{s,max} \leq \alpha_s \\ F\dfrac{(\alpha_s - \alpha_s^{min})^r}{(\alpha_{s,max} - \alpha_s)^s} & \alpha_s^{min} < \alpha_s < \alpha_{s,max} \\ 0 & \alpha_s \leq \alpha_s^{min} \end{cases}$$

where $F=0.1$, $r=2$, $s=5$, $\alpha_s^{min} = 0.1$

Various expressions had been proposed for the normal frictional stress p_c (Johnson and Jackson, 1987;Johnson et al., 1990; Ocone et al., 1993). The expression proposed by Johnson and Jackson (1987) has been widely adopted and was used in present simulations. In Johnson and Jackson's equation, frictional interaction between particles does not occur at α_s <0.5(α_s^{min} = 0.5). In fact both frictional and kinetic stresses exist at 0.1<α_s <0.5. Thus α_s^{min} = 0.5 is not appropriate especially for dense phase pneumatic conveying in horizontal pipe. In this paper α_s^{min} is equal to 0.1. Slug flow could be divided into three parts: dilute regime, intermediate regime and dense regime. Dilute regime locates at the upper part of the pipe and dense regime locates at the lower part of the pipe. Traditionally, the dilute flow in the so-called rapid granular flow regime is modeled based on the classic kinetic theory (Gidaspow, 1994), while the dense flow in the so-called quasi-static regime is modeled based on frictional stress model (Srivastava and Sundaresan, 2003). In the former the particles experience short and fast collisions, while in the latter the particles experience enduring contacts giving rise to frictional stress. The intermediate regime is expected to exist at the interface between those regimes, where the mass and momentum of particle phase exchange violently. Both kinetic stress and frictional stress play an important role. Therefore frictional stress cannot be omitted in the intermediate regime. In dilute regime, the particles cannot enduringly contact so that frictional stress could be omitted compared with kinetic stress. The expression for the normal frictional stress proposed by Johnson and Jackson (1987) is a semi-empirical equation and F, r, sare material dependent constants. Constant F is adjusted by comparing the experimental data with the pressure gradients predicted.

Turbulence Model

In the kinetic theory of granular flow, the granular temperature describes the particle–particle collisions and is introduced as a measure of single particle velocity fluctuation. Actually, particle cluster is one of the most significant features in dense phase pneumatic conveying. Thus it is necessary to consider both small-scale fluctuation due to particle–particle collision and large-scale particle fluctuation due to particle turbulence (Pu et al., 2008). Gas/particle turbulent kinetic energy and turbulent kinetic energy dissipation rate equation (i=gas, solid, $l\neq i$) are given by

$$\frac{\partial}{\partial t}(\alpha_i\rho_i k_i) + \nabla \cdot (\alpha_i\rho_i \mathbf{U}_i k_i) = \nabla \cdot \left(\alpha_i \frac{\mu_{i,t}}{\sigma_k} \nabla k_i\right) + (\alpha_i G_{k,i} - \alpha_i\rho_i\varepsilon_i)$$

$$+ \beta(C_{li}k_l - C_{il}k_i) - \beta(\mathbf{U}_l - \mathbf{U}_i)\frac{\mu_{l,t}}{\alpha_l\sigma_l}\nabla\alpha_l + \beta(\mathbf{U}_l - \mathbf{U}_i)\frac{\mu_{i,t}}{\alpha_i\sigma_i}\nabla\alpha_i \tag{11}$$

$$\frac{\partial}{\partial t}(\alpha_i\rho_i\varepsilon_i) + \nabla \cdot (\alpha_i\rho_i \mathbf{U}_i\varepsilon_i) = \nabla \cdot \left(\alpha_i \frac{\mu_{i,t}}{\sigma_\varepsilon} \nabla\varepsilon_i\right) + \frac{\varepsilon_i}{k_i}(C_{1\varepsilon}\alpha_i G_{k,i} - C_{2\varepsilon}\alpha_i\rho_i\varepsilon_i)$$

$$+ C_{3\varepsilon}\frac{\varepsilon_i}{k_i}\left[\beta(C_{li}k_l - C_{il}k_i) - \beta(\mathbf{U}_l - \mathbf{U}_i)\frac{\mu_{l,t}}{\alpha_l\sigma_l}\nabla\alpha_l + \beta(\mathbf{U}_l - \mathbf{U}_i)\frac{\mu_{i,t}}{\alpha_i\sigma_i}\nabla\alpha_i)\right] \tag{12}$$

where U_i is the phase-weighted velocity, and $G_{k,i}$ is the generation of turbulent kinetic energy. The gas or particle phase turbulent viscosity is defined as

$$\mu_{t,i} = \rho_i C_\mu \frac{k_i^2}{\varepsilon_i} \tag{13}$$

The constants in these equations are $C_\mu = 0.09$, $C_{1\varepsilon} = 1.42$, $C_{2\varepsilon} = 1.68$, $C_{3\varepsilon} = 1.2$.

Boundary conditions and calculation method

The boundary conditions for gas phase and particle phase applied in the simulations are given as follows:

Gas phase boundary conditions:

At the inlet, a non-uniform axial velocity distribution was specified:

$$v_{gz,in}(r) = \frac{60}{49}\frac{U_g}{1 - \alpha_{s,in}}(1 - 2r/D)^{1/7} \tag{14}$$

$$U_g = \frac{G_g}{\pi D^2/4} \tag{15}$$

where U_g is the superficial velocity, G_g is the gas flow rate, and $\alpha_{s,in}$ is the solid volume fraction at inlet.

Turbulent energy:

$$k_{g,in} = 0.004v^2_{g,in}$$

$$(16)$$

Dissipation rate of turbulent energy:

$$\varepsilon_{g,in} = 2k^{3/4}_{g,in}/(\kappa D)$$

$$(17)$$

where κ=0.4187, D is the inside diameter of the pipe.

At the outlet, the gas-phase pressure was ambient and all other variables were subjected to the Newmann boundary conditions $\partial\varphi/\partial z=0(\varphi=v_g, k_g, \varepsilon_g)$.

At the wall a no sli p boundary condition was used and the standard wall functions were specified for the gas phase.

Particle phase boundary conditions:

At the inlet, uniform distributions in axial velocity, solid granular temperature, and solid volume fraction were provided.

$$v_{sz,in} = \frac{M_s}{\alpha_{s,in}\rho_s\pi D^2/4}$$

$$(18)$$

$$\alpha_{s,in} = \frac{M_s}{\rho_s}\left(\frac{M_g}{\rho_g} + \frac{M_s}{\rho_s}\right)^{-1}$$

$$(19)$$

$$k_{s,in} = 0.004v^2_{s,in}$$

$$(20)$$

$$k_{s,in} = 0.004v_{s,in}^2$$

(21)

$$\Theta_{s,in} = 0.004v_{s,in}^2$$

(22)

where M_s is the particle mass flow rate and M_g is the gas mass flow rate.

At the outlet all other variables were subjected to the Newmann boundary conditions $\partial\varphi/\partial z=0$ ($\varphi=v_s$, k_s, ε_s,Θ_s).

At the wall, the partial slip boundary conditions proposed by Johnson and Jackson (1987) [20] were applied in the present study.

$$\tau_{sw} = \frac{\sqrt{3}}{6}\pi\rho_s g_{0,ss}\phi' \frac{\alpha_s}{\alpha_{s,max}}\sqrt{\Theta_s}\mathbf{v}_{sw}+\mu p_s$$

(23)

$$q_w = \frac{\sqrt{3}}{6}\pi\rho_s g_{0,ss}\varphi \frac{\alpha_s}{\alpha_{s,max}}\sqrt{\Theta_s}|\mathbf{v}_{sw}|^2 -\frac{\sqrt{3}}{4}\pi\rho_s g_{0,ss}(1-e_w^2)\frac{\alpha_s}{\alpha_{s,max}}\Theta_s^{3/2}$$

(24)

where $'$ is the specularity coefficient representing the fraction of total momentum transferred to the wall when particle collides with it. q_w is the flux of granular temperature toward the wall; μ is the particle–wall friction coefficient; and e_w is the restitution coefficient of particle–wall collision. As an initial condition, an empty pipe is assumed.

The set of coupled conservation equations were discretized into systems of linear equations using the finite-volume method. The power-law interpolation scheme of discretization was used for momentum solutions, which provides solutions with accuracy between that obtained from first- and second-order schemes. Compared with other higher-order schemes, this method was more robust and less computationally intensive. Taking advantage of the strong coupling between pressure and velocity, the phase-coupled SIMPLE iterative algorithm was used for the pressure-velocity coupling.

Under relaxation factors were necessary to obtain convergent solutions with the iterative scheme used. Relaxation factors used in the current study had typical values of 0.2 for all the variables. The solution is assumed to converge when the sum of normalized residuals has fallen below a specified level δ.

$$\frac{\sum_{CV}|R_\Phi|}{\sum_{CV}|\Phi|} < \delta$$

(25)

where R_Φ is the local residual of the Φ equation, Φ is the corresponding local quantity and subscript CV denotes the control volume. The time step used is 1×10^{-6} s. In the present study, δ was assigned as 0.001. The simulations were carried out in supercomputer "Sunway-2000A" in Wuxi Supercomputing Center using software package Fluent. In the present study, the length of horizontal pipe is 5 m with the inside diameter of 10 mm.

RESULTS AND DISCUSSION

Comparison of Experiment Data with Prediction of Pressure Gradient

Fig. 3 shows the comparison of the predicted pressure gradient with the experimental data at two transport pressures. It is obvious that a good agreement between them is reached. It can be seen that when the solid mass flow rate is kept constant and as the superficial velocity increases, the pressure gradient first decreases, then goes through a minimum point, and subsequently starts to increase. The resistance property shown in Fig. 3(a) is not comprehensive because it is difficult to keep the gas flow rate constant with high level at high pressure in experiments. But the resistance property shown in Fig. 3(b) is very clear. The predicted tendency is similar to that of the experiment. At low superficial velocity, the pulverized coal deposits in the pipe with high settled layer. The friction between the settled layer and wall is the main contributor to the pressure gradient. The settled layer becomes thinner with increasing superficial velocity, so the pressure gradient decreases. At high superficial velocity, the particles are assumed to fly with a uniform distribution over the pipe cross-section. This state is referred to as fully suspended flow. In this conveying state, the pressure gradient increases with increasing superficial velocity, because the friction between the gas and wall increases with U_g^2. There is no sharp transition from suspended flow to slug flow. The transition is gradual. So there exists a point where the pressure gradient is the minimum.

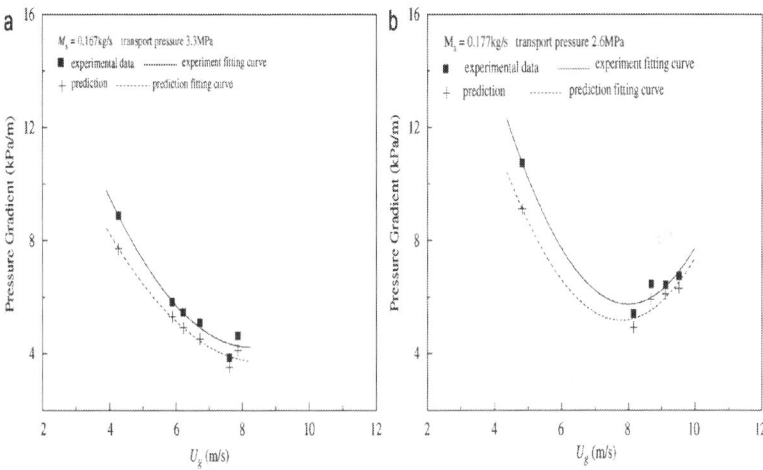

Figure 3: Pressure gradient vs. superficial velocity: (a) transport pressure 3.3 MPa and (b) transport pressure 2.6 MPa.

Solid Volume Fraction Distribution at Cross Section

Fig. 4a shows the experimental ECT image representing cross sectional solid volume fraction distribution in the pipe, and Fig. 4b shows the simulated solid volume fraction contours. The parameters used in simulations are listed in Table 3. The ECT image shows that the pulverized coal layer deposits in the pipe and the top surface of the settled layer takes on a distinct concave. the particles above the settled layer are transported in a disperse state. Jaworski and Dyakowski (2002) and Zhu et al. (2004) applied ECT systems to study dense phase pneumatic conveying in horizontal pipe and inclined riser, respectively. From their ECT images, it can be found that the top surface of the settled layer had the same shape. The simulation results agreed well with the ECT image. According to the prediction of particle concentration in Fig. 4b, the cross section can be divided into three parts: the dilute regime (a), the intermediate regime (b) and the dense regime (c). The dilute regime (a) locates at the upper part of the pipe, where the particles experience short and fast collisions. The dense regime (c) locates at the lower part of the pipe, where the particle concentration is high and the particles experience enduring contacts giving rise to frictional stress. For horizontal pneumatic conveying in a confined volume, the intermediate regime (b) is typically expected to exist at the interface between the dense regime and the dilute regime, where both collision and frictional stresses might be present. It is commonly believed that there exists fierce mass and momentum exchange of particle phase both between the intermediate regime and the dilute regime and between the intermediate regime and the dense regime. Some particles in the dilute regime enter into the intermediate regime due to gravity, at the same time some particles in the intermediate regime enter into the dilute regime due to particle–particle collisions or dragging by gas. The same case happens between the intermediate regime and the dense regime. It also can be founded that there are no clear boundaries among these regimes. It indicates that the particle movement in the pipe is very complex and affected by many factors. Fig. 4c shows the simulated solid volume fraction contours at central vertical plane. From the picture, it can be seen that the flow type predicted belongs to stratified flow.

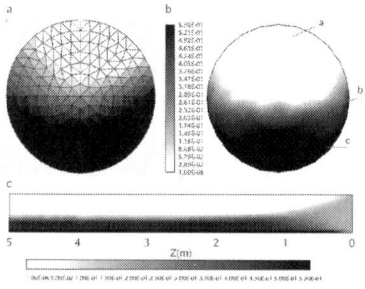

Figure 4: Solid volume fraction distribution at cross section and central vertical plane (a) ECT image of solid volume fraction, (b) solid volume fraction

contours simulated and (c) solid volume fraction contours simulated at central vertical plane.

Table 3: Experimental conditions and parameters used in simulations

Description	Symbol	Values	Units
Sending hopper pressure	Psend	3.6	MPa
Receiving hopper pressure	Prec	2.8	MPa
Solid mass flow rate	Ms	0.227	kg/s
Pipe inside diameter	D	10	
Superficial velocity	Ug	6.37	mm
Particle diameter	ds	80	m/s
Particle–particle restitution coefficient	ess	0.8	μm
Particle–wall restitution coefficient	ew	0.6	dimensionless
Specularity coefficient	Φ'	0.02	dimensionless
Angle of internal friction	Φ	28.5	dimensionless
Maximum solid fraction	αs,max	0.55	dimensionless

Fig. 5 shows the predicted solid volume fraction contours at different superficial velocity. Here the solid mass flow rate is 0.177 kg/s with the particle diameter of 37 μm, the superficial gas velocity U_g is 4.82 m/s, 8.15 m/s and 9.52 m/s, respectively. It is clearly observed that as the superficial velocity increases, the settled layer becomes thinner, which fits common sense. Two factors may contribute to the phenomenon. For the solid mass flow rate constant, the particle velocity increases with the superficial velocity increasing. It leads to the solid volume fraction decrease. On the other hand, the gas velocity fluctuations become stronger at higher superficial velocity, which enhances the mass and momentum exchange of particle phase between the dilute regime and dense regime and leads to the settled layer decrease.

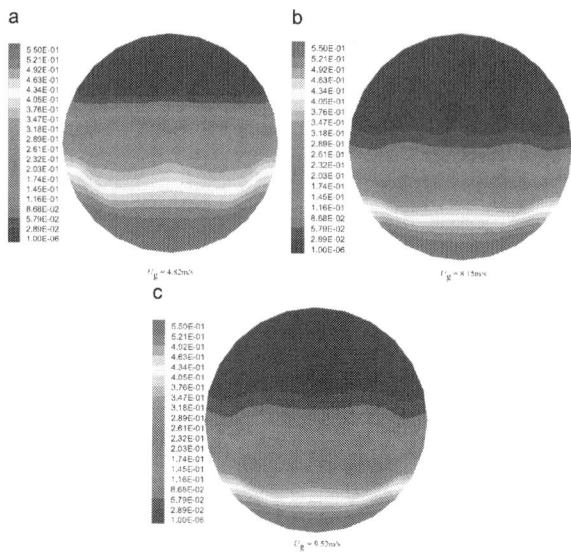

Figure 5: Prediction of solid volume fraction at different superficial velocity: (a) U_g=4.82 m/s, (b) U_g=8.15 m/s and (c) U_g=9.52 m/s.

The Formation and Motion Process of Slug Flow

Although the number of experimental studies describing the characteristics of the flow field has grown over the last few years, only partial information has been collected and published. As a result, it is impossible to comprehensively validate the predictions of the model, and so it is decided to compare the similarity between the characteristics of the predicted flow field and the experimentally determined flow fields. Thus, a qualitative comparison is presented.

Single Slug Flow of Pulverized Coal

The prediction of the single slug movement along a 5 m horizontal pipe is presented in Fig. 7. Here the solid mass flow rate is 0.167 kg/s with a diameter of 37 μm, the superficial gas velocity U_g is 4.26 m/s and the solid volume concentration is 29.3%. Fig. 6 shows the solid feeding rate changing with time.

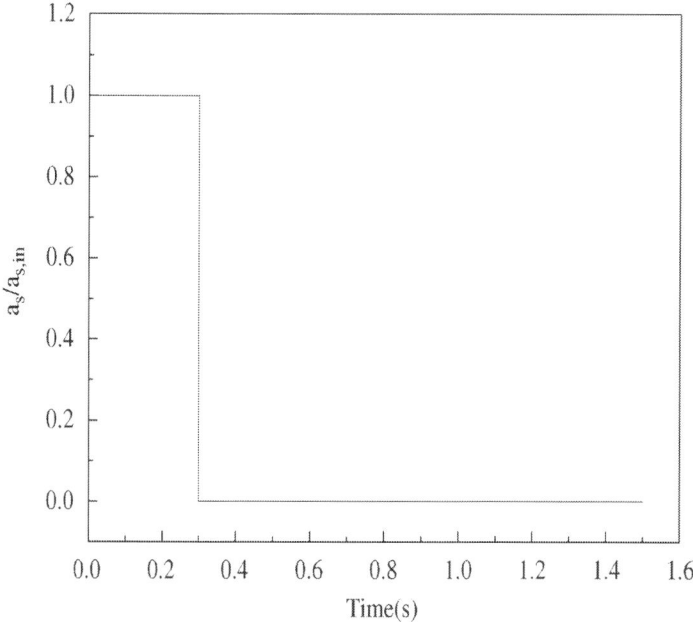

Figure 6: Solid volume fraction at inlet vs. time.

Pulverized coal is fed into the pipe inlet during a time interval of 0.3 s. During that period, a long slug with high particle layer is formed near the pipe inlet in Fig. 7. After the first 0.3 s, only gas flow enters into the pipe inlet. Thus, a long slug built artificially during the first 0.3 s of the numerical simulation is formed. As time proceeds, the slug moves from the right to the left (t=0.3–0.7 s). Finally the slug moves away from the pipe (t=0.7–1.0 s). In addition, moving slug leaves a thinner layer behind it and short afterwards the layer is carried away by gas until the pipe becomes empty (t=1.1–1.5 s).

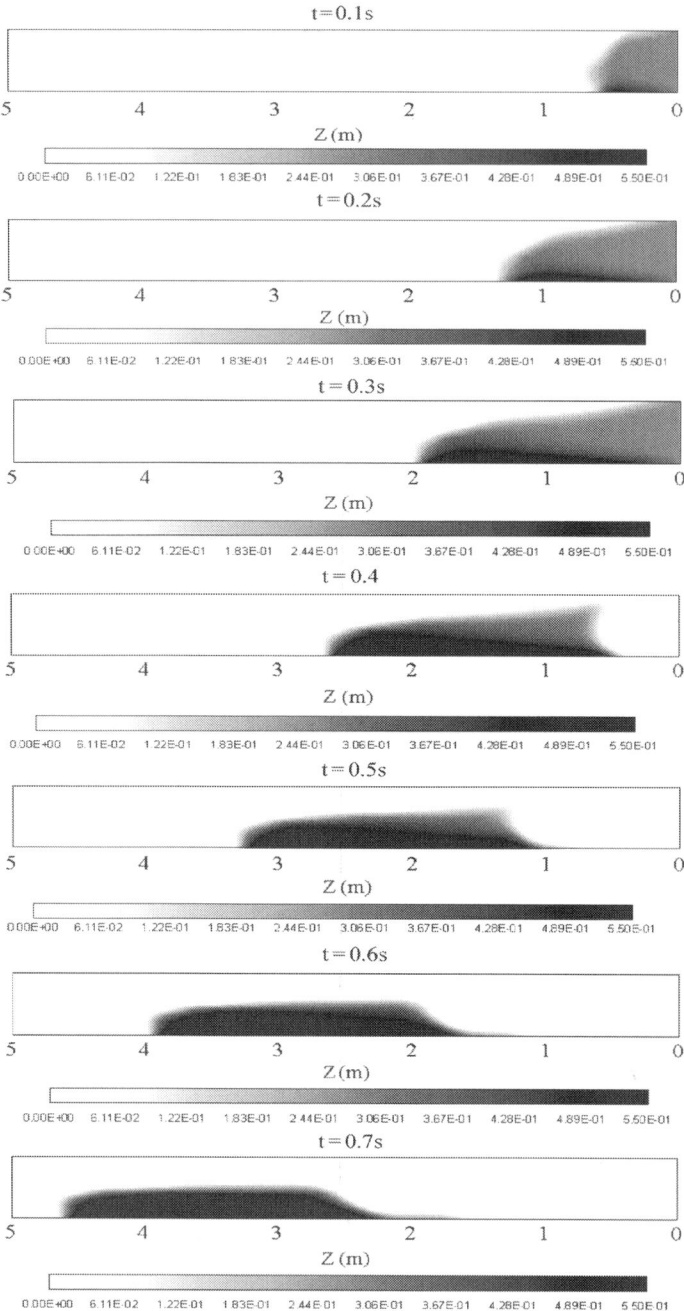

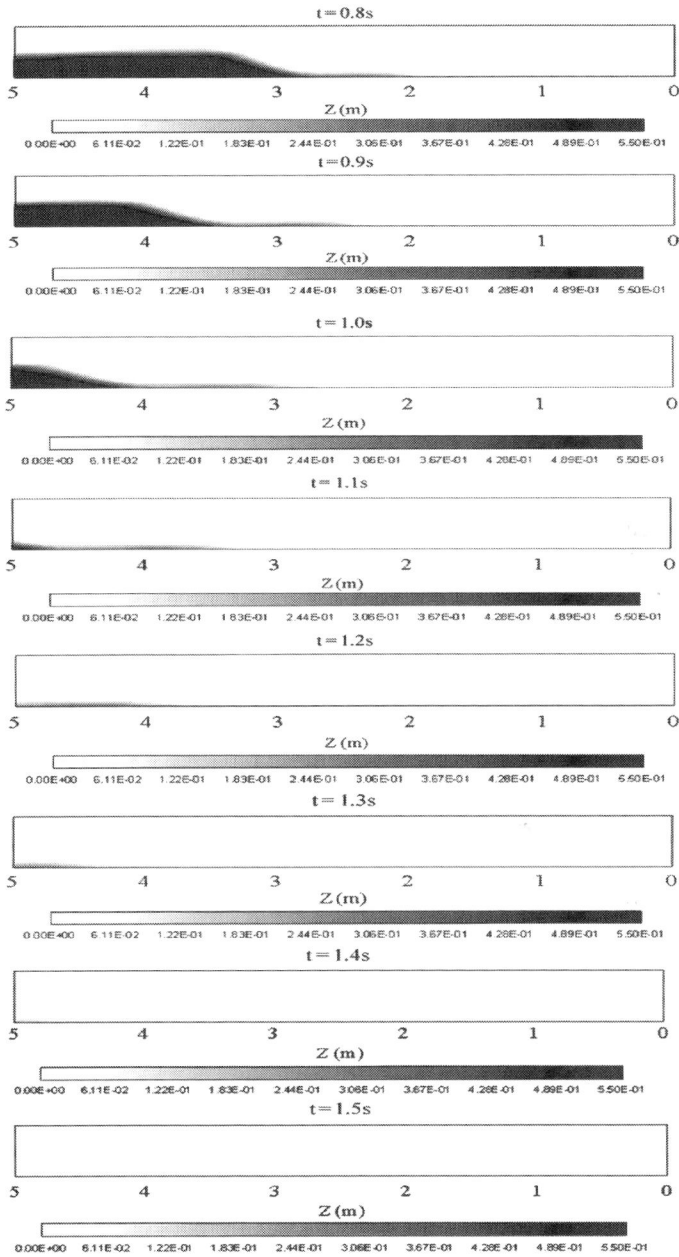

Figure 7: Prediction of single slug movement over central vertical plane.

Fig. 8 shows the pressure history at $Z=2500$ mm. The pressure at $Z=2500$ mm keeps constant until the slug does not arrive ($t=0–0.4$ s). When the forehead of the slug arrives the pressure begins to increase ($t=0.4–0.7$ s), then the pressure reaches a constant value when the tail of the slug leaves ($t=0.7–0.8$ s). After the slug arrives at the outlet of the pipe, the pressure begins to decrease with the outflow of the slug ($t=0.8–1.5$ s). The simulation results are similar to the description of the pressure change of single slug by Fan (2002).

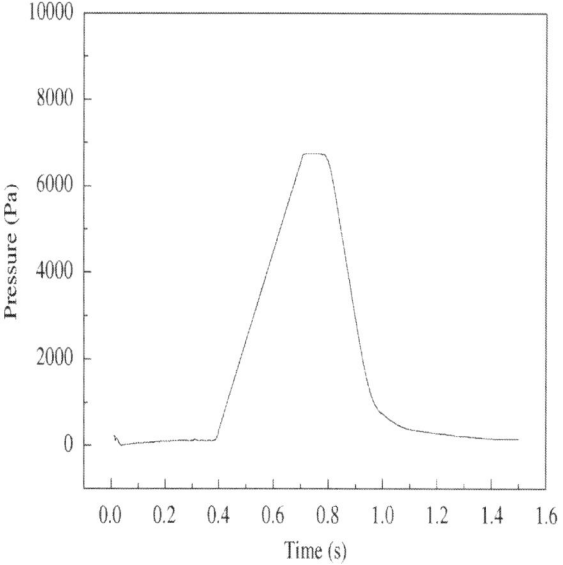

Figure 8: Pressure vs. time at $Z=2500$ mm.

During the initial visualization test it was difficult to get the clear photographs in the horizontal visualization section for the pulverized coal with a diameter of 37 μm because of its strong wall adhesion. In order to eliminate the particle–wall adhesion, the pulverized coal with a diameter of 300 μm was used. Fig. 9 shows a series of seven photographs illustrating a passage of single slug through the visualization section in the horizontal pipe. By comparing Fig. 9 with Fig. 7, it can be seen that the simulation results are similar to the characteristics of the experimental flow fields. As the slug passes through the pipe gradually, there is a thin layer left behind, as seen in Fig. 9 ($t=0.096$ s) which is similar to the prediction results in Fig. 7$t=1.1$ s.

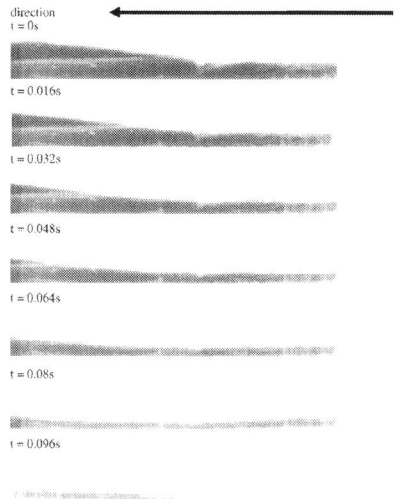

Figure 9: Visualization photos of single slug flow in the horizontal pipe with high speed camera.

Slug Flow of Pulverized Coal

To illustrate the continuous slug flow, the simulation of slug flow was carried out. The simulation conditions are the same as the single slug flow. The pulsed solid feeding rate was given in Fig. 10.

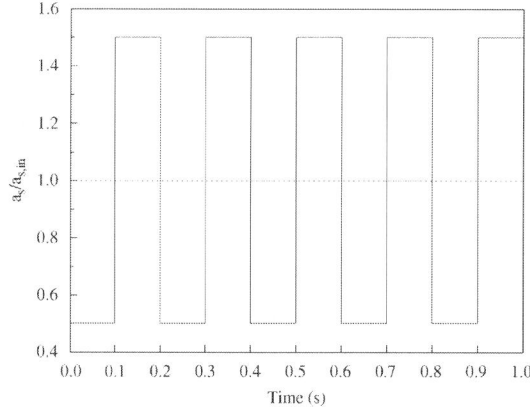

Figure 10: Solid volume fraction at inlet vs. time.

Fig. 11 shows side view of the predicted slug flow pattern. As an initial condition, the pipe is empty. The flow is from right to left. It is clearly observed that the slug moves like wave motion.

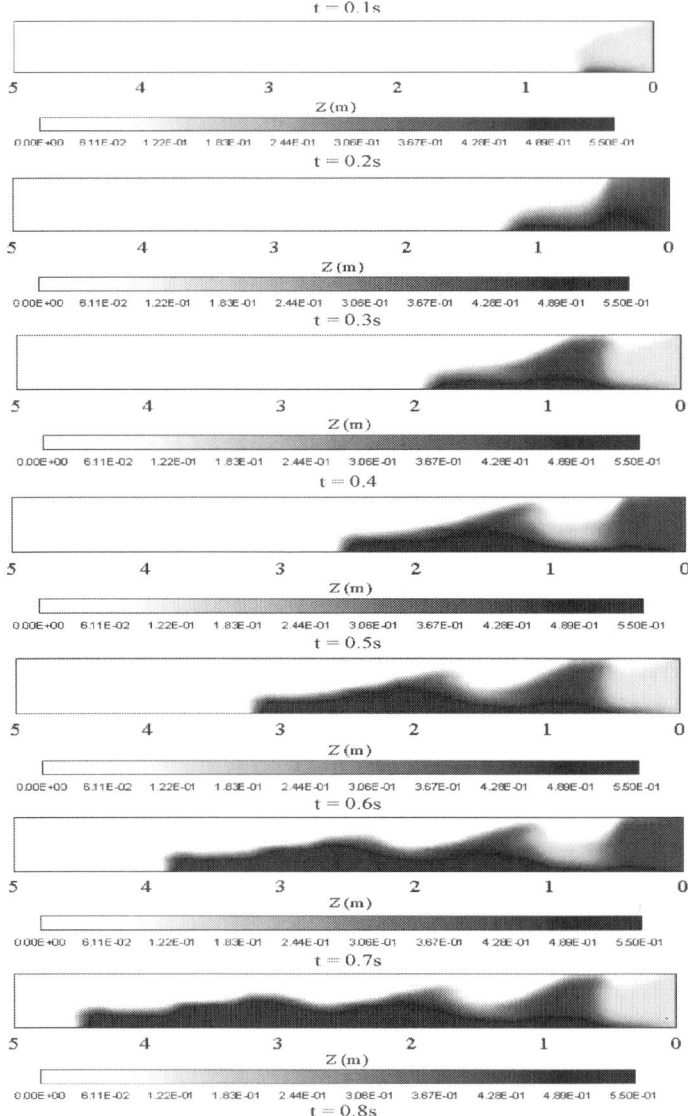

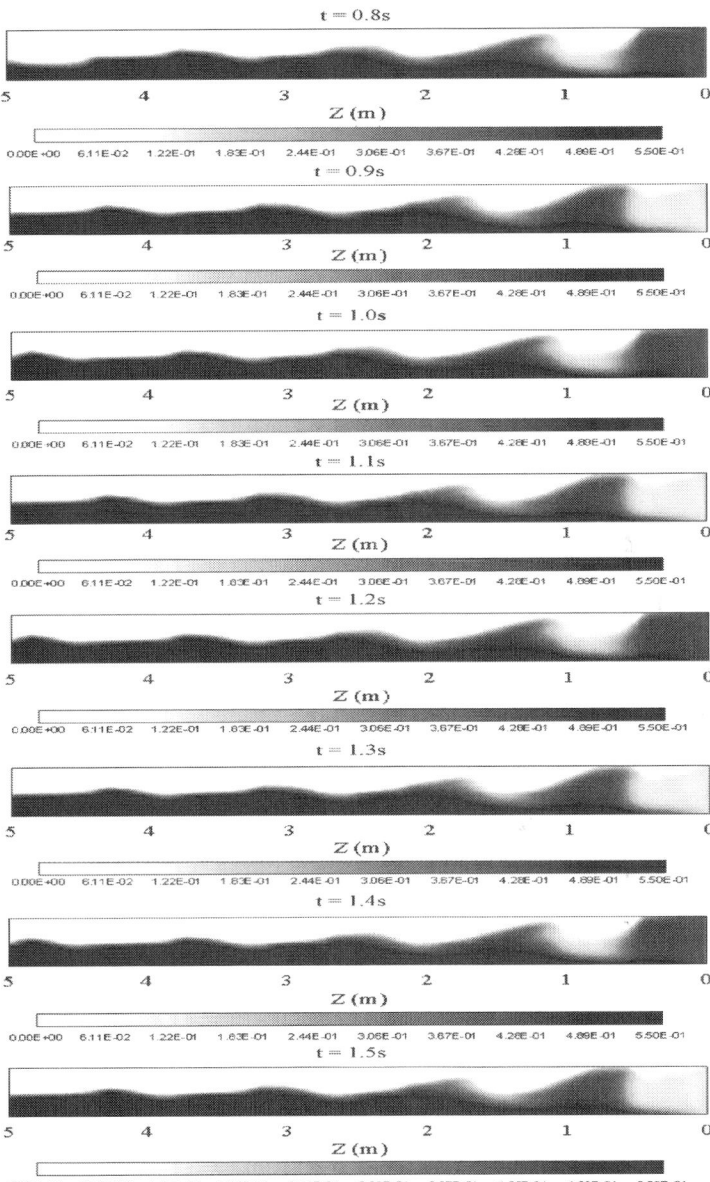

Figure 11: Prediction of slug flow movement over central vertical plane.

Fig. 12 shows a set of four photographs illustrating a passage of slug through the visualization section in the horizontal pipe.

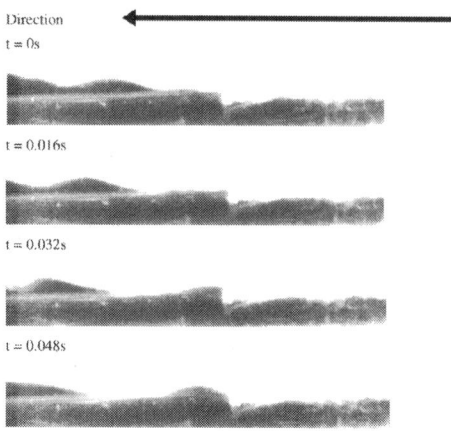

Figure 12: High speed camera visualization of slug flow in the horizontal pipe.

Due to electrostatic force and the wall adhesion, it is difficult to obtain the clear photographs for the fine pulverized coal (37 μm). The simulation of slug flow was carried out. Here the solid mass flow rate is 0.214 kg/s with a diameter of 300 μm, the superficial gas velocity U_g is 5.92 m/s and the solid volume concentration is 25.45%. Fig. 13 shows side view of the predicted results. From Fig. 13, it is obvious that the flow proceeds in the form of wave and the flow pattern in Fig. 13(a) is very realistic. The simulation results inFig. 13(b) resemble the flow as the photographs obtained by high speed camera in the experiments.

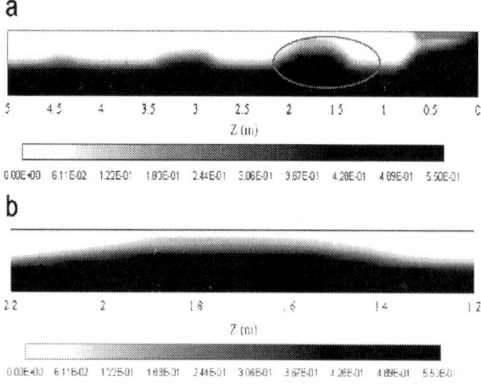

Figure 13: Prediction of slug flow movement over central vertical plane.

CONCLUSIONS

The hydrodynamics of dense phase pneumatic conveying of pulverized coal in horizontal pipe was predicted employing a three dimensional model in which a kinetic–frictional stress model for particle phase was included. This model assumed that the frictional and kinetic stresses are additive. A normal frictional stress model of Johnson and Jackson (1987) and a modified frictional shear viscosity model of Syamlal et al. (1993) were used for friction stress. The gas phase and particle phase were treated as turbulent flow, respectively. The experiment was carried out to validate the prediction results. The solid concentration distribution at cross section was obtained and compared with the measurement by ECT. Then the effects of superficial velocity on solid concentration distribution were discussed. The formation and motion process of slug flow was simulated and compared with the photographs by high speed camera. Simulation results are in good agreement with experimental data. Based on the presented results, the following conclusions may be drawn:

- As the superficial velocity increases pressure gradient decreases first and then increases. The model predictions of the pressure gradient are in good agreement with the experimental data.

- In dense phase pneumatic conveying in horizontal pipe at high pressure, the pulverized coal deposits in the pipe and the top surface of the settled layer has a distinct concave. The predictions are validated by ECT image.

- The formation and motion process of slug flow is demonstrated and the slug moves like wave motion, which is similar to the photographs obtained by high speed camera in the experiments.

ACKNOWLEDGMENTS

The authors gratefully acknowledge the financial support from the National Key Program of Basic Research in China 2004CB217702 and the financial support from natural science fund for colleges and universities in Jiangsu Province 09KJD610003. The authors are grateful to Dr. Yang Daoye for the cooperation of the project.

REFERENCES

1. Benyahia, S., Arastoopour, H., Knowlton, T.M., Massah, H., 2000. Simulation of particles and gas flow behaviour in the riser section of a circulating fluidized bed using the kinetic theory approach for the particulate phase. Powder Technology 12, 24–33.

2. Chen, X.P., Fan, C.L., Liang, C., Pu, W.H., Lu, P., Zhao, C.S., 2007. Investigation on characteristics of pulverized coal dense-phase pneumatic conveying under high pressure. Korean Journal of Chemical Engineering 24 (3), 499–502.

3. Curtis, J.S., van Wachem, B., 2004. Modeling particle-laden flows: a research outlook. A.I.Ch.E. Journal 50 (11), 2638–2645.

4. Fan, C., 2002. Plug flow dense phase pneumatic conveying. Advances in Mechanics 32, 599–612.

5. Geldart, D., Ling, S.J., 1990. Dense phase conveying of fine coal at high total pressure. Powder Technology 62, 243–252.

6. Gidaspow, D., 1994. Multiphase Flow and Fluidization: Continuum and Kinetic Theory Descriptions. Academic Press, New York.

7. Gidaspow, D., Jung, J., Raj, K.S., 2004. Hydrodynamics of fluidization using kinetic theory: an emerging paradigm 2002 Flour-Daniel lecture. Powder Technology 148, 123–141. Huilin, L., Yurong, H., Wentie, L., Ding, J., Gidaspow, D., Bouillard, J., 2004. Computer simulations of gas–solid flow in spouted beds using kinetic– frictional stress model of granular flow. Chemical Engineering Science 59, 865–878.

8. Jaworski, A.J., Dyakowski, T., 2002. Investigations of flow instabilities within the dense pneumatic conveying system. Powder Technology 125, 279–291.

9. Johnson, P.C., Jackson, R., 1987. Frictional-collisional constitutive relations for granular materials, with application to plane shearing. Journal of Fluid Mechanics 176, 67–93.

10. Johnson, P.C., Nott, P., Jackson, R., 1990. Frictional–collisional equations of motion for particulate flows and their application to chutes. Journal of Fluid Mechanics 210, 501–535.

11. Konrad, K., 1986. Dense-phase pneumatic conveying: a review. Powder Technology 49, 1–35.

12. Levy, A., 2000. Two-fluid approach for plug flow simulations in horizontal pneumatic conveying. Powder Technology 112, 263–272.

13. Liang, C., Zhao, C.S., Chen, X.P., Pu, W.H., Lu, P., Fan, C.L., 2007. Flow characteristics and Shannon entropy analysis of dense-phase pneumatic conveying of pulverized coal with variable moisture content at high pressure. Chemical Engineering and Technology 30, 926–931.

14. Lun, C.K., Savage, S.B., Jeffrey, D.J., 1984. Kinetic theories for granular flow: inelastic particles in coquette flow and slightly inelastic particles in a general flow field. ournal of Fluid Mechanics 140, 223–256.

15. Makkawi, Y., Ocone, R., 2006. A model for gas–solid flow in a horizontal duct with a smooth merge of rapid-intermediate-dense flows. Chemical Engineering Science 61, 4271–4281.

16. Makkawi, Y.T., Wright, P.C., Ocone, R., 2006. The effect of friction and inter-particle cohesive forces on the hydrodynamics of gas–solid flow: a comparative analysis of theoretical predictions and experiments. Powder Technology 163, 69–79.

17. McKeen, T., Pugsley, T., 2003. Simulation and experimental validation of freely bubbling bed of FCC catalyst. Powder Technology 129, 139–152.

18. Ocone, R., Sundaresan, S., Jackson, R., 1993. Gas–particle flow in a duct of arbitrary inclination with particle–particle interaction. A.I.Ch.E. Journal 39, 1261–1271.

19. Patil, D.J., van Sint Annaland, M., Kuipers, J.A.M., 2005a. Critical comparison of hydrodynamic models for gas–solid fluidized beds—Part I: bubbling gas–solid fluidized beds operated with a jet. Chemical Engineering Science 60, 57–72.

20. Patil, D.J., van Sint Annaland, M., Kuipers, J.A.M., 2005b. Critical comparison of hydrodynamic models for gas–solid fluidized beds—Part II: freely bubbling gas–solid fluidized beds. Chemical Engineering Science 60, 73–84.

21. Pu, W.H., Zhao, C.S., Xiong, Y.Q., Liang, C., Chen, X.P., Lu, P., Fan, C.L., 2008. Threedimensional numerical simulation of dense pneumatic conveying of pulverized coal in a vertical pipe at high pressure. Chemical Engineering and Technology 31, 215–223.

22. Sakai, M., Koshizuka, S., 2009. Large-scale discrete element modeling in pneumatic conveying. Chemical Engineering Science 64, 533–539.

23. Savage, S.B., 1998. Analyses of slow high-concentration flows of granular materials. Journal of Fluid Mechanics 377, 1–26.

24. Schaeffer, D.G., 1987. Instability in the evolution of equations describing incompressible granular flow. Journal of Differential Equations 66, 19–50.

25. Sinclair, J.L., Jackson, R., 1989. Gas-particle flow in a vertical pipe with particle– particle interactions. A.I.Ch.E. Journal 35, 1473–1486.

26. Srivastava, A., 2001. Dense-phase gas–solid flows in circulating fluidized beds. Ph.D. Thesis, Deptartment of Chemical Engineering, Princeton University.

27. Srivastava, A., Sundaresan, S., 2003. Analysis of a frictional-kinetic model for gas– particle flow. Powder Technology 129, 72–85.

28. Sundaresan, S., 2000. Modeling the hydrodynamics of multiphase flow reactors: current status and challenges. A.I.Ch.E. Journal 46, 1102–1105.

29. Syamlal, M., Rogers, W.A., O'Brien, T.J., 1993. MFIX documentation and theory guide, DOE/METC94/1004, NTIS/DE94000087. Electronically available from /http://www.mfix.orgS.

30. Tardos, G.I., 1997. A fluid mechanistic approach to slow, frictional flow of powders. Powder Technology 92, 61–74.

31. Tardos, G.I., McNamara, S., Talu, I., 2003. Slow and intermediate flow of frictional bulk powder in the coquette geometry. Powder Technology 131, 23–39.

32. Tsuji, T., Yabumoto, K., Tanaka, T., 2008. Spontaneous structures in threedimensional bubbling gas-fluidized bed by parallel DEM-CFD coupling simulation. Powder Technology 184, 132–140.

33. Tsuji, Y., Tanaka, T., Ishida, T., 1992. Lagrangian numerical simulation of plug flow of cohesionless particles in a horizontal pipe. Powder Technology 71, 239–250.

34. Van Wachem, B.G.M., Schouten, J.C., Van den Bleek, C.M., 2001. Comparative analysis of CFD models of dense gas–solid systems. A.I.Ch.E. Journal 47, 1035–1051.

35. Van Wachem, B.G.M., Almstedt, A.E., 2003. Methods for multiphase computational fluid dynamics. Chemical Engineering Journal 96, 81–98.

36. Yang, D.Y., 2008. Study on electrical capacitance tomography with thick pipeline in gas/solid two-phase flow. Ph.D. Thesis, Southeast University.

37. Zheng, Y., Wang, X.T., Zhen, Q., Wei, F., Jin, Y., 2001. Numerical simulation of the gas–particle turbulent flow in riser reactor based on k-e-kp-ep two-fluid model.

38. Chemical Engineering Science 56, 6813–6822.

39. Zhu, H.P., Zhou, Z.Y., Yang, R.Y., Yu, A.B., 2008. Discrete particle simulation of particulate systems: a review of major applications and findings. Chemical Engineering Science 63, 5728–5770.

40. Zhu, H.P., Zhou, Z.Y., Yang, R.Y., Yu, A.B., 2007. Discrete particle simulation of particulate systems: theoretical developments. Chemical Engineering Science 62, 3378–3392.

41. Zhu, K.W., Wong, C.K., Rao, S.M., Wang, C.H., 2004. Pneumatic conveying of granular solids in horizontal and inclined pipes. A.I.Ch.E. Journal 50, 1729–1745.

CFD Investigation of the Pipe Transport of Coarse Solids in Laminar Power Law Fluids

M. Eesa, M. Barigou

School of Chemical Engineering, The University of Birmingham, Edgbaston, Birmingham B15 2TT, UK

ABSTRACT

A numerical parametric study of the laminar pipe transport of coarse particles in non-Newtonian carrier fluids of the power law type has been conducted using an Eulerian–Eulerian computational fluid dynamics (CFD) model. The predicted flow fields have been successfully validated by experimental measurements of particle velocity profiles obtained using a positron emission particle tracking technique, whilst solid–liquid pressure drop has been validated using relevant correlations gleaned from the literature. The study is concerned with nearly-neutrally buoyant particles flowing in a horizontal or vertical pipe. The effects of various parameters on the flow properties of such mixtures have been investigated over a wide range of conditions. The variables studied are:

particle diameter (2–9 mm), mean solids concentration (5–40% v/v), mean mixture velocity (25–125 mm s^{-1}), and rheological properties of the carrier fluid (k=0.15–20 Pa sn;n=0.6–0.9). A few additional runs have been conducted for shear thickening fluids, i.e. n>1. Whilst the effects of varying the power law parameters and the mixture flowrate for shear thinning fluids are relatively small over the range of values considered, particle size and solids concentration have a significant bearing on the flow regime, the uniformity of the normalised particle radial distribution and of the normalised velocity profiles of both phases, and the magnitude of the solid–liquid pressure drop. The maximum particle velocity is always significantly less than twice the mean flow velocity for shear thinning fluids, but it can exceed this value in shear thickening fluids. In vertical down-flow, particles are uniformly distributed over the pipe cross-section, and particle diameter and concentration have little effect on the normalised velocity and concentration profiles. Pressure drop, however, is greatly influenced by particle concentration.

INTRODUCTION

Particle transport through pipes is an important operation in many industries including foods, pharmaceuticals, chemicals, oil, mining, construction and power generation industries. In many of these applications the carrier fluid may be highly viscous and may have a non-Newtonian rheology and flow is usually laminar. The transport of particles in laminar non-Newtonian fluids offers certain advantages including: a maximum apparent viscosity at the centre of the pipe in a shear thinning fluid which tends to assist particle suspension, though this effect may be damped by particle radial migration and enhanced settling velocities in sheared fluids; a minimum apparent fluid viscosity at the pipe wall which reduces pressure drop and makes it increase relatively slowly as a function of mixture flowrate; and when the fluid exhibits an apparent yield stress, this tends to aid the suspension of coarse particles in the central region of the pipe (Chhabra and Richardson, 1999).

Such systems have not been widely studied in the literature as the vast majority of the documented data relate to water-based slurries of fine particles. There is, therefore, a clear need for experimental data and models to describe the flow of large particles in non-Newtonian fluids as they are relevant to a number of industrial applications such as the conveying of particulate food mixtures, gravel, and coal lumps. These are heterogeneous mixtures of complex rheology such that the assumption of a continuum, as usually used in the modelling of fine pseudo-homogeneous suspensions, is clearly inapplicable. The use of non-Newtonian carrier fluids, however, has been hampered by a lack of understanding of the complex solid–liquid flows that result, and has only been addressed in a handful of studies (including, Charles and Charles, 1971; Ghosh and Shook, 1990; Duckworth et al., 1983 and Duckworth et al., 1986; Chhabra and Richardson, 1985; Fairhurst et al., 2001; Barigou et al., 2003; Gradeck et al., 2005; Legrand et al., 2007;Eesa and Barigou, 2008). Detailed measurements of the flow field and pressure drop in these systems are scarce. Some limited studies experimented with magnetic resonance imaging (McCarthy et al., 1996) and ultrasound Doppler velocimetry (Guer et al., 2003). Barigou et al. (2003) used positron emission particle tracking (PEPT) to study the flow of coarse (d=5 or 10 mm) nearly neutrally-buoyant alginate particles in shear-thinning carboxymethylcellulose (CMC) fluids and reported information on flow regimes, solid phase velocity profiles, and particle passage time distributions.

Computational modelling in this specific area has been even more limited. In a rare attempt, Krampa-Morlu et al. (2004) used computational fluid dynamics (CFD) to study the flow features of coarse aqueous solid–liquid slurries in turbulent upward flow including velocity profiles. The model, implemented using the commercial CFD software CFX 4.4 (ANSYS Inc.), was tested using experimental data from Sumner et al. (1990). The particles had a density of 2650 kg m^{-3} and a diameter of 0.47 or 1.7 mm and were simulated at concentrations up to 30% v/v. The authors concluded that, using the default settings, the code failed to accurately predict important features of the flow.

Recently, we investigated the capabilities of CFD to model the flow of coarse nearly-neutrally buoyant particles in shear-thinning CMC fluids in a horizontal pipe for a small number of flow cases (Eesa and Barigou, 2008). CFD results of particle velocity profiles were validated using experimental data obtained by PEPT, while pressure drop predictions were compared with a number of selected correlations from the literature. In this paper, a CFD model, based on the commercial code ANSYS CFX 10.0, is used to conduct a detailed parametric study of the transport of nearly neutrally-buoyant coarse particles in laminar non-Newtonian flow in both horizontal and vertical flow. Sample validation data for particle velocity profiles and pressure drop are presented. The aim of this work is to describe in detail the role of the various physical properties and process parameters and to demonstrate the capability of CFD in modelling such complex flows.

FORMULATION OF CFD MODEL

The Eulerian–Eulerian two-fluid model was adopted here, whereby both the liquid and solid phases are considered as continua. Treating the solid phase as an Eulerian phase is possible provided that the inter-phase interactions are adequately modelled. In fact, the Eulerian approach has been reported to be efficient at simulating multiphase flows once the interaction terms are included (see for example, Hu et al., 2001). The Eulerian–Lagrangian model, however, which simulates the solid phase as a discrete phase and so allows particle tracking, is in principle more realistic. However, after evaluating the relevant literature as well as conducting a number of simulation trials, it was concluded that the number of dispersed particles that can be tracked within all the different commercial CFD software available is currently very limited, thus restricting the applicability of the Eulerian–Lagrangian model to only dilute mixtures well below ~5% v/v (van Wachem and Almstedt, 2003).

The laminar flow of coarse particles in a non-Newtonian fluid is assumed to be governed by the following equations which form the basis of the Eulerian–Eulerian CFD model used.

Continuity Equations

Assuming isothermal flow, the continuity equations can be written for both the liquid and solid phase as follows (van Wachem and Almstedt, 2003)

$$\frac{\partial}{\partial t}(\rho_f C_f) + \nabla \cdot (\rho_f C_f U_f) = 0$$

$$(1)$$

$$\frac{\partial}{\partial t}(\rho_s C_s) + \nabla \cdot (\rho_s C_s U_s) = 0$$

(2)

with the constraint

$$C_f + C_s = 1$$

(3)

where the subscripts f and s denote, respectively, the fluid and solid phase, C is the mean volume fraction, ρ is density, U is velocity vector, and t is time.

Momentum Equations

The momentum equation for each phase is written so that it includes, along with all the forces acting on that phase, an inter-phase momentum transfer term that models the interaction between the two phases (Enwald et al., 1996; van Wachem and Almstedt, 2003); thus for the liquid phase and for the solid phase

$$\rho_f C_f \left[\frac{\partial U_f}{\partial t} + U_f \cdot \nabla U_f \right] = -C_f \nabla P + C_f \nabla \cdot \bar{\bar{\tau}}_f + C_f \rho_f g - M$$

(4)

$$\rho_s C_s \left[\frac{\partial U_s}{\partial t} + U_s \cdot \nabla U_s \right] = -C_s \nabla P + C_s \nabla \cdot \bar{\bar{\tau}}_s$$
$$+ C_s \rho_s g - \nabla P_s + M$$

(5)

where P is pressure, g is the gravitational acceleration vector, $\bar{\bar{\tau}}$ is the viscous stress tensor, P_s is solid pressure, and M is the interfacial momentum transfer per unit volume made up of the drag force, F_d, and the lift force, F_l . The forces on the right-hand side of the momentum equations are the pressure force, viscous force, gravitational force, as well as the particle-particle interaction force for the solid phase represented by the solid pressure term (∇_{Ps}). As the inter-particle interactions increase with solids concentration, the inclusion of this term is particularly important for highly concentrated suspensions (i.e., where $C_s > 0.2$). This solid pressure term is, therefore, a function of the solids concentration (Gidaspow, 1994), thus

$$P_S = P_S(C_S)$$

$$(6)$$

and therefore,

$$\nabla P_S = G(C_S)\nabla C_S$$

$$(7)$$

The function G (C_s) is referred to as the elasticity modulus, and is expressed as follows

$$G(C_S) = G_0 e^{E(C_S - C_{sm})}$$

$$(8)$$

where G_0 is the reference elasticity modulus, E is the compaction modulus, and C_{sm} is the maximum packing parameter (maximum solid loading). There are no generally accepted values for these parameters; however, the values $G_0=1$ Pa, $E=20–600$, have been suggested by Bouillard et al. (1989). The maximum packing parameter C_{sm} was determined by Thomas (1965) as 0.625 for spherical particles.

Inter-Phase Drag Force

The inter-phase drag force per unit volume, F_d, is expressed as

$$F_d = \frac{3C_D}{4d} C_S \rho_f |U_S - U_f|(U_f - U_S)$$

$$(9)$$

where C_D is the drag coefficient of a single particle and d is the particle diameter (van Wachem and Almstedt, 2003; Kleinstreuer, 2003). For Reynolds numbers outside Stokes flow, the drag coefficient is a function of the flow regime and can be estimated using one of the many empirical correlations which exist (Lareo et al., 1997). For concentrated suspensions up to $C_s=0.2$, the expression by Wen and Yu (1966) can be used to calculate C_D, thus

$$F_d = \frac{3C_D}{4d} C_S \rho_f |U_S - U_f|(U_f - U_S)$$

$$(10)$$

where $Re'=(1-_{Cs})R_{ep}$, and

$$C_D = (1 - C_s)^{-1.65} \max \left[\frac{24}{Re'}(1 + 0.15Re'^{0.687}), \, 0.44 \right]$$

(11)

where Re_p is the generalised particle Reynolds number which depends on the fluid effective viscosity, $\mu_{f'}$ and the local slip velocity between the two phases, represented here by the sedimentation velocity, $u_{\infty'}$ of the particle in an infinite expanse of the carrier fluid (Brown and Heywood, 1991).

For higher solids concentrations (i.e., $C_s > 0.2$), the Gidaspow drag model can be used where the inter-phase drag force per unit volume is given by (Ding and Gidaspow, 1990)

$$F_d = \left[150 \frac{C_s^2 \mu_f}{C_f d^2} + \frac{7C_s \rho_f |U_s - U_f|}{4d} \right] (U_f - U_s)$$

(12)

Lift Force

The lift force exerted by the velocity gradient is given by

$$F_l = C_s \rho_f C_l (U_s - U_f) \times (\nabla \times U_f)$$

(13)

where C_l is the lift coefficient (van Wachem and Almstedt, 2003). A wide range of values for C_l can be found in the literature; the choice of the C_l value used in the simulations is addressed below with other simulation parameters (Section 3.2).

Suspension Viscosity

The presence of solid particles in a carrier fluid affects the shear rate distribution, and hence the suspension viscosity. The apparent viscosity of the suspension, $\mu_{susp'}$ can therefore be written in terms of the apparent viscosity of the fluid, $\mu_{a'}$ and a solids viscosity, $\mu_{s'}$ thus

$$\mu_{susp} = (1 - C_s)\mu_a + C_s \mu_s$$

(14)

The suspension viscosity is usually expressed in terms of the relative viscosity, μ_r, defined as

$$\mu_r = \frac{\mu_{susp}}{\mu_a}$$

(15)

The well-known Einstein equation for estimating the relative viscosity of very dilute suspensions with negligible particle-particle interactions is given by (Chakrabandhu and Singh, 2005)

$$\mu_r = \frac{\mu_{susp}}{\mu_a} = 1 + 2.5C_s$$

(16)

For more concentrated suspensions, however, inter-particle interactions must be taken into account and modifications made to Einstein's equation resulted in a number of expanded versions of it, including the following expression due to Thomas (1965)

$$\mu_r = 1 + 2.5C_s + 10.05C_s^2 + 20.84C_s^3$$

(17)

where the third-order term accounts for particle-particle interactions. The constants in Eq. (17) have been determined mainly for fine particles in Newtonian carrier fluids. However, their use has been validated for coarse particles in non-Newtonian fluids by Chakrabandhu and Singh (2005). After evaluation of various expressions using their own experimental data, these authors concluded that Eq.(17) provided the closest estimates over the entire range of experimental variables considered in their study.

Carrier Fluid Viscosity

The fluid rheology implemented in the CFD model is described by the power law model, thus

$$\mu_a = k\dot{\gamma}^{n-1}$$

(18)

where $\dot{\gamma}$ is shear rate, k is the consistency index and n is the flow behaviour index, with n<1 for a shear-thinning fluid, n>1 for a shear-thickening fluid, and n=1 for a Newtonian fluid. The volumetric flowrate, Q, of a power law fluid in a pipe is given by the following exact equation (Chhabra and Richardson, 1999)

$$Q = \frac{n\pi R^3}{(3n+1)} \left(\frac{R}{2k} \cdot \frac{\Delta p}{L} \right)^{1/n}$$

(19)

where R is the pipe radius, and P/L is the pressure drop per unit length. The fluid velocity profile can also be derived, thus

$$u(r) = \frac{n}{n+1} \left(\frac{\Delta P.R}{2kL} \right)^{1/n} R \left[1 - \left(\frac{r}{R} \right)^{(n+1)/n} \right]$$

(20)

where r is radial position.

DESCRIPTION OF CFD SIMULATIONS

Simulations were set up in three dimensions because of the axial asymmetry of the two-phase flow using the commercial software package ANSYS Workbench 10.0, and were solved using the CFX-Solver component of the package. The geometry used consisted of a pipe of diameter D=45 mm. The pipe length,L, was much greater than the maximum entrance length, L_e, required for flow to fully develop. In single-phase Newtonian laminar flow, L_e can be estimated from (Shook and Roco, 1991)

$$\frac{L_e}{D} = 0.062 Re_t$$

(21)

where $R_{et} = \rho_f u_m D / \mu_f$ is the tube Reynolds number and u_m is the mean flow velocity. There is no correlation available for estimating L_e in two-phase solid–liquid flow, however, but for the almost neutrally-buoyant particles used here, Eq. (21) should give a reasonable estimate of L_e.

In fact, the use of a shear thinning fluid and the presence of solid particles—as discussed below—both lead to flatter velocity profiles and, hence, Eq. (21) is even likely to yield conservative estimates. Whilst such estimates were used as a guide, a series of numerical trials were conducted using different pipe lengths. For all of the cases simulated here, a pipe length of 600 mm was sufficient to give a fully developed solid–liquid flow through most of the pipe length whilst keeping computational cost low. Using a longer pipe did not affect the results. The geometry was meshed into approximately $1.8{\times}10^5$ tetrahedral cells. The 3D grid was optimised through a mesh-independence study. Inflation layers covering about 20% of the pipe radius were created near the pipe wall in order to accurately account for the locally high parameter gradients in that region.

Simulation of Single-Phase Fluid Flow

Simulations of the carrier fluid flowing alone were performed first to serve both as an initial validation of the code and the numerical grid, and to reveal the effects of solid particles on the liquid velocity profile by comparing the velocity profile of the liquid flowing alone to that which exists in a solid–liquid suspension. The power law viscosity relationship (Eq. (18)) was used in the CFD model, and the values of the power law parameters investigated are shown in Table 1. A mass flowrate boundary condition was used at the pipe inlet, while static pressure was specified at the outlet. The usual no-slip boundary condition was adopted at the pipe wall.

Table 1: Range of parametric CFD study for shear thinning fluids

Direction of flow	Parameter investigated	k (Pa sⁿ)	n(−)	ρ_f (kg m⁻³)	ρ_s (kg m⁻³)	d(mm)	C_s(−)	u_m (mm s⁻¹)	$Re_L = \dfrac{\rho f^{u_n}}{\mu_f}$	$Re_L = \dfrac{\rho f^{u^{2-n}_0}}{\mu_f}$
Horizontal	d	0.16	0.81	1000	1020	2−9	0.25	125	60	0.003–0.37
	C_s	0.16	0.81	1000	1020	4	0.05 – 0.40	125	60	0.026
	u_m	0.16	0.81	1000	1020	4	0.25	25–125	9–60	0.020–0.026
	n	0.15	0.6 – 0.9	1000	1020	4	0.30	125	50–116	0.025–0.031
	k	0.15 – 20	0.65	1000	1020	4	0.30	125	1–102	0–0.026
Vertical	d	0.16	0.81	1000	1020	2–8	0.25	66	28	0.003–0.22
	C_s	0.16	0.81	1000	1020	4	0.05 – 0.30	66	28	0.023

The advection terms in the governing momentum equation (Eq. (4) with C_f=1 and M=0 for single-phase flow) were discretised using a second-order differencing scheme, which is more accurate than a first-order scheme (Shaw, 1992). In the finite volume method used for discretising the momentum equation, the variable value at an integration point, φ_{ip}, is calculated from its value at the upwind node, φ_{up} , and the variable gradient, $\nabla\varphi$, thus

$$\phi_{ip} = \phi_{up} + \beta\nabla\phi.\Delta\vec{r}$$

(22)

where β is a blend factor and $\Delta\vec{r}$ is the vector from the upwind node to the integration point. The numerical scheme employed here was the numerical advection correction scheme, wherein a constant value of β is specified and the variable gradient is calculated as the average of the adjacent nodal gradients. With β=0, the scheme is first-order accurate; in the current simulation, β was set equal to 1 which is second-order accurate. The solution was assumed to have converged when the mass and momentum residuals reached 10^{-5} for all of the equations solved. This typically required ~100 iterations.

Simulation of Two-Phase Solid–Liquid Flow

As discussed above, the solid particles were introduced in the continuous liquid phase as a second Eulerian phase. The viscosity of the solid phase was modelled using Eq. (14) with a relative suspension viscosity as given by Eq. (17). Details of the numerical simulations conducted are summarised in Table 1.

The boundary conditions at inlet and outlet of the pipe were the mixture mass flowrate and the exit static pressure, respectively. The homogeneous volumetric fraction of each phase was specified at the inlet. Using flowrate as a boundary condition is the common way of formulating pipe flow problems, i.e. one designs a system to deliver a given flowrate. Note, however, that using a pressure-specified inlet boundary condition is a stricter way of testing the CFD code as a flowrate boundary condition may be perceived as a way of helping to steer the simulation towards the right solution. This pressure option was tested but it did not affect the results of the CFD computations. Two different wall conditions were used for the liquid and solid phases.

For the liquid phase the usual no-slip condition was used, while for the solid-phase free-slip was assumed in order to prevent the solid phase from adhering to the wall, which is consistent with the real flow behaviour of coarse particles near a solid boundary and is customary practice in two-phase flow modelling.

Due to the complexity of the solid–liquid flows considered here, simulations initially required a great deal of experimentation and optimisation. Of primary importance was the appropriate modelling of forces and interactions between the two phases. The buoyancy force was taken into account by the density difference between the liquid and solid. Forces due to inter-particle collisions required the introduction of an additional "solid pressure" term into the solid phase momentum equation (Eq. (5)). A solid pressure model based on the Gidaspow model (Eqs. (6), (7) and (8)) was used with default values of the model parameters: G_0=1 Pa,E=600, and C_{sm}=0.625. The drag force was modelled using the Wen–Yu drag model (Eq. (10)) for solid concentrations up to 20% v/v, and the Gidaspow drag model (Eq. (12)) for higher concentrations. It should be noted here that the particle Reynolds number (Eq. (11)) was computed based on the local apparent viscosity of the fluid used, thus taking account of the non-Newtonian behaviour of the fluid (He et al., 2001). The lift force was modelled using Eq. (13) with the default value C_l=0.5; a range of C_l values were tested but they did not affect the results.

The so-called "high resolution scheme" was implemented in discretising the advection terms in the governing equations. In this scheme, the value of the blend factor, β, in Eq. (22), is not constant but is calculated locally to be as close to 1 as possible without resulting in non-physical parameter values. This scheme is therefore intended to satisfy the requirements of both accuracy and boundedness. Imposing a second-order accurate scheme (i.e. β=1) in such complex simulations may result in difficulty of convergence.

Numerical convergence under steady state mode was too difficult to achieve for solid–liquid flows and, consequently, the simulations were run in the transient mode. In steady state simulations, and due to the absence of time-dependence, the fluid acceleration is not modelled in the same way as it physically occurs. It is usually recommended that simulations of steady state nature should be run transiently when convergence becomes difficult, in order to enhance the stability of convergence. The modelling of time variation smooths out the way in which the solution changes from one iteration to the next (Shaw, 1992). A small time step of 0.05 s was used to ease convergence. Eventually, the solution reached a steady state and fulfilled the convergence criterion which was set at a residual target RMS=10^{-4} for all of the equations solved. On average, 200 time steps were required, with 1–6 iterations to achieve convergence for all of the equations at each time step.

VALIDATION OF CFD SIMULATIONS

As pointed out above, detailed measurements of the flow field in solid–liquid suspensions are very scarce due to the lack of suitable measurement techniques. In our earlier work (Barigou et al., 2003; Fairhurst et al., 2001; Fairhurst, 1998) we carried out extensive experiments using the technique of PEPT to determine the trajectories and velocity profile of coarse solid particles flowing in non-Newtonian CMC fluids. These unique sets of experimental results were used here to validate the CFD simulations. CFD predictions of pressure drop in solid–liquid flow, on the other hand, were assessed using appropriate correlations gleaned from the literature. The validation of velocity profile and pressure drop in single-phase fluid flow simulations, however, was based on exact analytical equations available in the literature.

Validation of Single-Phase Fluid Flow

The fluids considered corresponded to 0.5 and 0.8 wt% aqueous CMC solutions. The flow field of each liquid flowing alone was obtained numerically. The predicted flowrate was within 1% of the exact solution (Eq. (19)) for the 0.5% CMC solution, and better than 3% for the more viscous 0.8% CMC solution. The numerically obtained velocity profiles are compared in Fig. 1 to the exact analytical profiles for a power law fluid (Eq. (20)). The agreement between theory and CFD is excellent for both fluids.

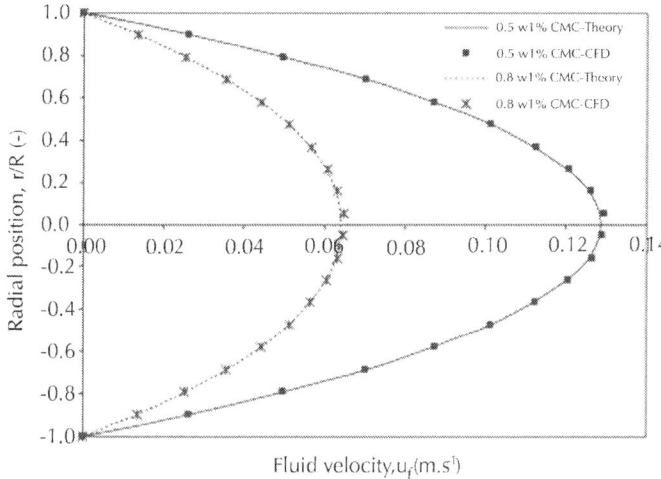

Figure 1: Comparison of theoretical and CFD velocity profiles for a power law fluid flowing alone: 0.5 wt% CMC (k=0.16 Pa sn, n=0.81),u_m=66 mm s^{-1}; 0.8 wt% CMC (k=0.75 Pa sn, n=0.71), u_m=33 mm s^{-1}.

PEPT Validation of Solid-Phase Velocity Profile

PEPT uses a single positron-emitting particle as a flow tracer which is tracked in 3D space and time within operating equipment to reveal its full Lagrangian trajectory. Details of the technique and its applications can be found in Barigou (2004) and Barigou et al. (2003). A gravity driven flow loop was used which consisted of a down pipe followed by a horizontal pipe both of 45 mm inner diameter. The solid particles were alginate spheres of 5 and 10 mm diameter. Experiments were performed for outlet solids volumetric concentrations up to 40% v/v and mixture velocities up to 125 mm s^{-1}. A 600 µm resin bead, containing the positron emitting radionuclide ^{18}F, was imbedded inside an alginate particle which was used as a radioactive tracer. The resin bead had no measurable effect on the density of the particle, and the alginate tracer thus had the same physical properties as any other particle and was truly representative of the other particles. During an experimental run, single tracers were injected into the loop and collected at the exit; at least 50 particle trajectories were measured in order to obtain a representative sample. The solid-phase velocity profile was constructed using an algorithm based on calculating the velocity of particles at different radial positions. This was achieved by following a particle at a particular radial position and timing its travel between successive locations. The velocity was then normalised by the mean mixture velocity, u_m, across the pipe. Sample results are presented in Fig. 2 and Fig. 3 for horizontal flow and vertical down-flow, respectively, alongside the CFD-predicted velocity profiles extracted at a section 100 mm upstream of the pipe exit. There is a close agreement between CFD and experiment with the simulated profiles generally falling well within the experimental error bars. In horizontal flow (Fig. 2), the solid phase velocity profile is asymmetrical about the central axis, with particles flowing faster near the top of the pipe cross-section than near the bottom. This is due to particle settling as discussed in more detail below (Section 5.1.1). In vertical down-flow (Fig. 3), however, the velocity profile is as would be expected symmetrical about the pipe centreline.

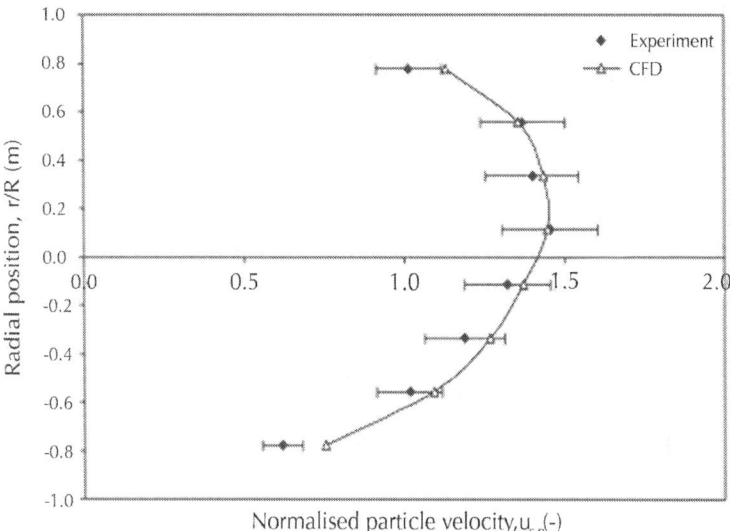

Figure 2: CFD-predicted and experimental solid-phase velocity profiles compared for horizontal flow: 0.5 wt% CMC (k=0.16 Pa sn, n=0.81);ρ_s=1020 kg m^{-3}; d=10 mm; C$_s$=0.30; u$_m$=65 mm s^{-1}.

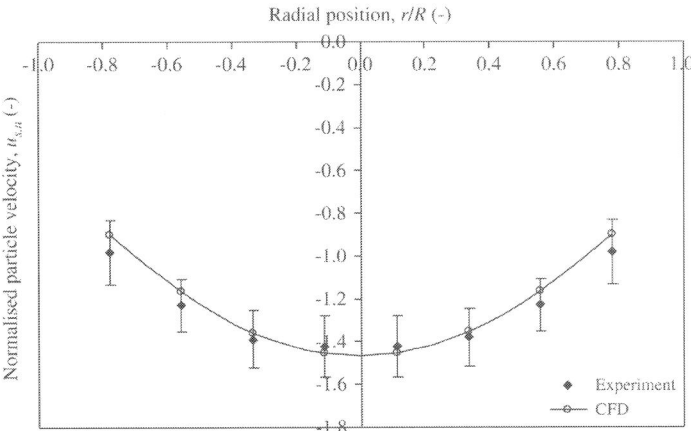

Figure 3: CFD-predicted and experimental solid-phase velocity profiles compared for vertical down-flow: 0.5 wt% CMC (k=0.16 Pa sn,n=0.81); ρ_s=1020 kg m^{-3}; d=10 mm; C$_s$=0.21; u$_m$=72 mm s^{-1}.

CFD simulation yields a smooth particle velocity profile because the Eulerian–Eulerian numerical model used treats the solid phase as a continuum rather than individual particles. However, such a model has proved capable of providing a good prediction of the solid phase velocity profile under a wide range of flow conditions including considerably viscous liquids and high solid concentrations up to 40% v/v. Validation results over a wider range of conditions can be found in Eesa and Barigou (2008).

Validation of Pressure Drop

One of the complications in estimating pressure drop in solid–liquid systems is the existence of different flow regimes which are dictated by the complex interaction of the many different variables involved. It is known that the presence of solid particles results in an increase in the pressure drop incurred, but there are no theoretical models available for calculating such a pressure drop. Furthermore, no experimental data could be found in the literature for the type of suspension considered here, i.e. coarse particles in non-Newtonian or even Newtonian laminar flow. However, many empirical or semi-empirical approaches do exist, and a number of relevant correlations gleaned from the literature were used to verify the CFD predictions (Durand, 1952; Zandi and Govatos, 1967; Newitt et al., 1955; Rasteiro et al., 1993; Gradeck et al., 2005).

CFD simulations were conducted using a range of particle diameters, solid concentrations and Reynolds numbers, as shown in Table 2. Full details of the validation are given in Eesa and Barigou (2008). Overall,Rasteiro et al.'s (1993) semi-empirical correlation provided the best agreement with CFD, as shown inTable 2. In most of the cases studied, the agreement was within ~15% even at high solid concentrations, which can be considered very good considering the complexity of the flow. Agreement between CFD andGradeck et al.'s (2005) correlation was also very good (<~15%) at low solid concentrations, but deteriorated at higher concentrations which were outside the range of their experimental data and therefore fell outside the correlation's range of validity (i.e. C_s>15% v/v).

Table 2: Comparison of CFD predictions of solid-liquid pressure drop with literature correlations

d(mm)	C_s(-)	$_s$ (kg m^{-3})	u_m (m s^{-1})	$Re_s = \frac{\rho f^{u_m D}}{\mu_f}$	P/L (Pa m^{-1})		
					CFD	Rasteiro et al. (1993) (Eq.(23))	Gradeck et al. (2005) (Eq. (25))
2	0.20	1020	0.221	100	624	629 (−1%)[a]	719 (−13%)[a]
	0.30	1020	0.221	100	848	815 (+4%)	1116 (−24%)
	0.40	1020	0.221	100	1070	1104 (−3%)	1712 (−38%)
5	0.20	1020	0.221	100	555	629 (−12%)	719 (−23%)
	0.30	1020	0.221	100	731	815 (−10%)	1116 (−34%)
5	0.10	1020	0.022	10	46	49 (−6%)	48 (−4%)
	0.10	1020	0.221	100	452	498 (−9%)	478 (−5%)
10	0.10	1020	0.022	10	42	49 (−15%)	48 (−12%)
	0.20	1020	0.221	100	488	629 (−22%)	719 (−32%)

Deviation of CFD prediction from correlation.

Rasteiro et al. (1993) considered the total pressure drop in a solid–liquid system to be made up of the kinetic energy loss, E_c, the viscous energy loss, E_v, and the energy loss due to particle-particle interactions,E_p, thus

$$\left[\frac{\Delta P}{L}\right]_{susp} = A_1 E_c + A_2 E_v + A_3 E_p$$

(23)

Where

$$E_c = \frac{\rho_f u_m^2}{d(1 - C_s)}; \quad E_v = \frac{\mu_f u_m}{(d(1 - C_s))^2}; \quad E_p = \frac{(\rho_s - \rho_f)g C_s^2}{(1 - C_s)^3}$$

(24)

A_1, A_2, and A_3 are empirical constants. Gradeck et al. (2005), on the other hand, used the analogy with single phase fluids to describe the pressure drop of dilute to moderately concentrated suspensions of nearly-neutrally buoyant coarse alginate particles in water, glucose solutions, and CMC solutions. The pressure drop was expressed in terms of the friction factor, f, where for laminar flow

$$f = \frac{16}{Re_{susp}}$$

(25)

and the suspension Reynolds number, Re_{susp}, is based on the mean suspension density, $_{susp}$, and the effective viscosity of the suspension, μ_{susp}.

Other older correlations (Durand, 1952; Zandi and Govatos, 1967; Newitt et al., 1955) showed much larger deviations from CFD, exceeding 100% in some cases. Their limitations in predicting the effects of solids concentration and particle size have also been demonstrated (Eesa and Barigou, 2008). The data presented here show that, overall, CFD is capable of giving good predictions of this important design parameter.

PARAMETRIC STUDY

A parametric study was conducted to determine the influence of various parameters on the flow properties of coarse solid particles in non-Newtonian fluids of the power law type. Most of the simulations were performed for shear thinning fluids (i.e. n<1) in horizontal flow, with a smaller set of vertical flow cases being carried out to investigate the influence of flow direction and these are discussed in a separate section. A few simulations were also conducted for shear thickening fluids (i.e. n>1). The variables investigated were: particle diameter, mean solids concentration, mean mixture velocity, and power law parameters. The range of the numerical experiments conducted for each parameter studied is summarised in Table 1. The results are discussed below in terms of the effects on solid and fluid phase velocity profiles, radial particle concentration profile, and pressure drop.

Particle Diameter

Particle diameter was varied in the range 2–9 mm, corresponding to particle-pipe diameter ratios ofd/D=0.044–0.200. Other flow parameters were held constant, as shown in Table 1.

Effect on Solid-Phase Velocity Profile

The velocity profile of the solid phase was obtained at a section 100 mm upstream of the pipe exit and was normalised using the mean mixture velocity, u_m. Sample results are shown in Fig. 4 for d=2, 4 and 9 mm. The solid phase velocity profile is generally asymmetrical about the central axis, with particles flowing at higher velocities near the top wall of the pipe than near the base, but the degree of asymmetry is dependent on particle diameter. At d=2 mm, the velocity profile is nearly symmetrical, with the particles close to the pipe wall constituting a slow moving annular region, while at higher diameters the asymmetry in the velocity profile becomes increasingly more pronounced. These features are consistent with our earlier experimental observations and PEPT measurements (Barigou et al., 2003; Fairhurst et al., 2001;Fairhurst, 1998). Such asymmetrical profiles have also been reported for sand and water slurries (Newitt et al., 1962) and for dilute food mixtures of up to 10% v/v solids (Fregert, 1995). The asymmetry in the solid phase velocity profile is a result of particle settling due to the density difference between the two phases, even though this difference here is relatively small. The CFD-predicted velocity profiles of nearly-neutrally buoyant particles (s= $_s$/ $_f$=1.02) and of neutrally buoyant particles (s=1.00) flowing under the same flow conditions are compared in Fig. 5. The velocity profile of neutrally buoyant particles is exactly symmetrical.

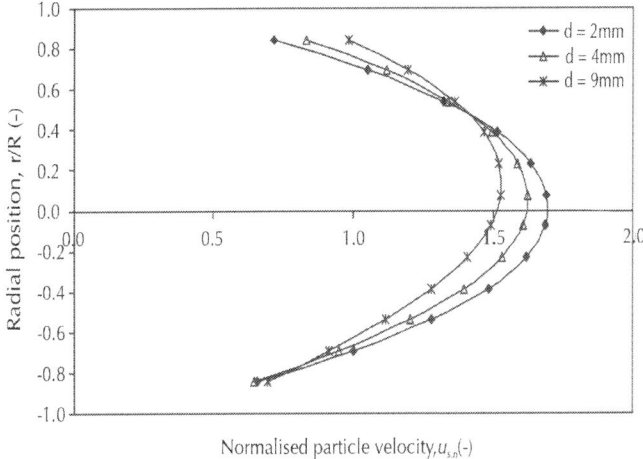

Figure 4: Effect of particle diameter on normalised solid-phase velocity profile: k=0.16 Pa sn; n=0.81; ρ_s=1020 kg m^{-3}; C$_s$=0.25;u$_m$=125 mm s^{-1}.

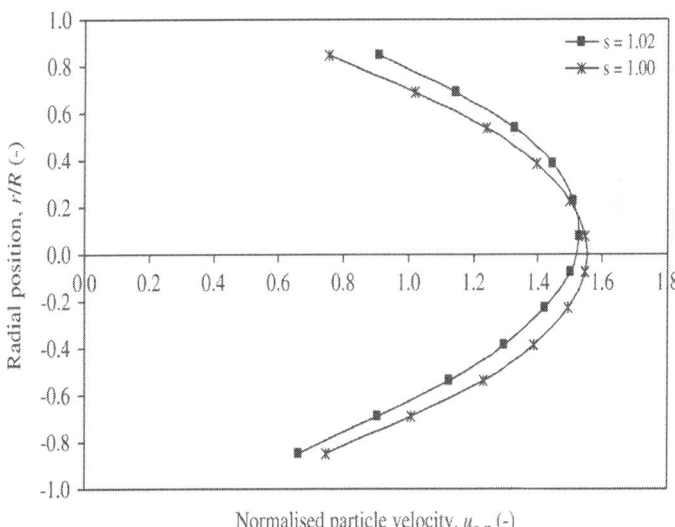

Figure 5: Normalised solid-phase velocity profile for neutrally buoyant and nearly-neutrally buoyant particles: k=0.16 Pa sn; n=0.81; d=7 mm;C$_s$=0.25; u$_m$=125 mm s^{-1}.

Two other features of the solid phase velocity profile which are affected by particle diameter are the position and value of the maximum velocity. While the maximum velocity occurs at the pipe centre for a fluid flowing alone, the maximum solid velocity in the solid–liquid mixtures studied lies generally above the centreline. The exact location of the fastest particle varies with particle diameter. For the smallest particles, the maximum of the velocity profile is located almost at the centre, but as d increases it moves above the centreline. Moreover, the value of the maximum particle velocity decreases for larger particles, which is in agreement with experimental observations for coarse particles of 5 and 10 mm diameter (Eesa and Barigou, 2008; McCarthy et al., 1997).

The particle velocity profile can have important implications in applications such as food processing where the aim is to safely sterilise the fastest particles without overcooking the slowest ones. The results in Fig. 4show that particles generally travel faster than the mean mixture velocity, u_m, and the fastest amongst them travel at a velocity considerably less than twice the mean value. It is common practice in food sterilisation, for example, to design holding tube lengths assuming that the maximum velocity is twice the average velocity (Lareo et al., 1997); such an assumption is clearly conservative for shear thinning fluids and can, therefore, result in losses of nutrients and quality. On the other hand, particles flowing in shear-thickening carrier fluids can reach a maximum velocity greater than twice the mean flow velocity, as shown in Fig. 6. It should be noted that a shear thickening fluid flowing alone has a maximum velocity greater than $2u_m$. At moderate values of n above one, such a velocity is reduced to below $2u_m$, as the velocity profile is flattened due to the presence of the particles, similar to what happens in shear thinning fluids (see Section 5.1.2below). At values of n substantially above one, both the fluid and particle maximum velocities can exceed $2u_m$. In such cases, estimation of a holding tube length based on a maximum velocity of twice the mean velocity will not be sufficient to ensure a safe process.

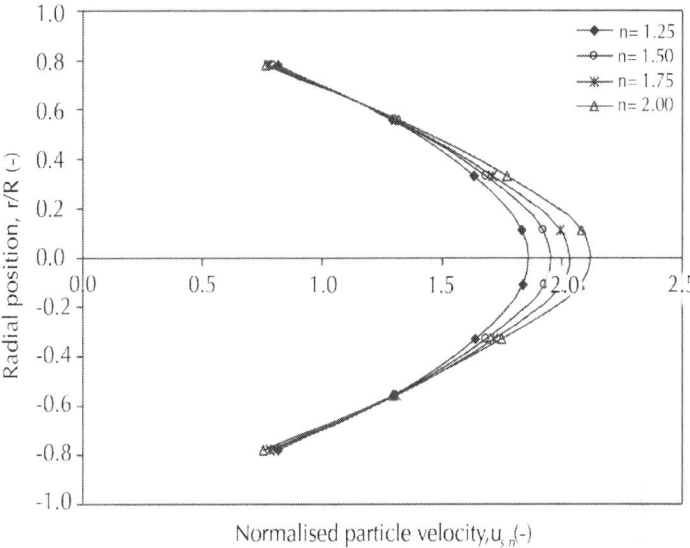

Figure 6: Normalised solid-phase velocity profile in shear-thickening carrier fluids: k=0.16 Pa sn; d=2 mm; ρ_s=1020 kg m^{-3}; C$_s$=0.30;u$_m$=125 mm s^{-1}.

Effect on Carrier Fluid Velocity Profile

Fig. 7 also shows how larger solid particles have a more significant blunting effect on the velocity profile of the fluid than smaller particles. In addition, particles induce a degree of asymmetry in the fluid velocity profile which increases with particle size. Fluid near the bottom of the pipe does not adhere to the wall despite the no-slip boundary condition imposed there. Solid particles near the bottom of the pipe have a non-zero velocity due to the solid free-slip condition at the wall. The flowing particles disrupt the boundary layer at the pipe wall and sweep the fluid away.

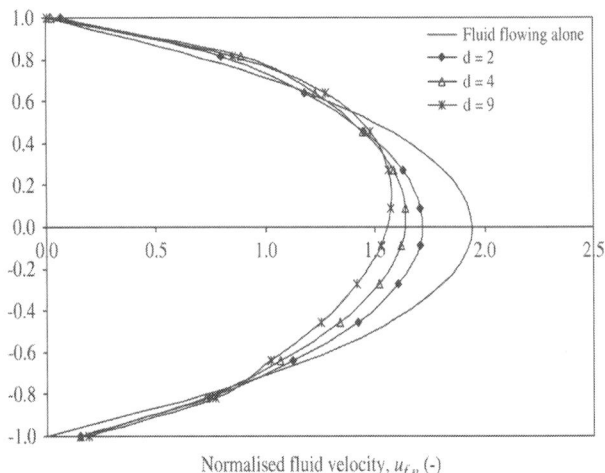

Figure 7: Effect of particle diameter on normalised carrier-fluid velocity profile: k=0.16 Pa sn; n=0.81; ρ_s=1020 kg m^{-3}; C$_s$=0.25;u$_m$=125 mm s^{-1}.

Effect on Particle Concentration Profile

The effect of particle size on the normalised radial distribution of the solid phase in the pipe is depicted inFig. 8. In general, three regions can be identified: (i) a low-concentration region near the top of the pipe cross-section; (ii) a central region with a nearly uniform solids concentration; and (iii) a high solids concentration region near the bottom of the pipe where the local solids fraction significantly exceeds the mean value. As discussed above, the effect of particle settling becomes more pronounced at larger particle diameters, leading to higher particle concentrations near the bottom of the pipe. Indeed, the CFD-predicted concentration profiles demonstrate this effect clearly in Fig. 8 for a range of particle sizes. Regions (i) and (iii) of the concentration profile at the top and bottom of the pipe, respectively, constitute a slow moving annular region encompassing a fast moving central core (region (ii)). The CFD results show that the diameter of the central core increases as particle size is reduced. Smaller particles, like neutrally buoyant particles, exhibit an entirely uniform radial distribution which is indicative of a pseudo-homogeneous flow.

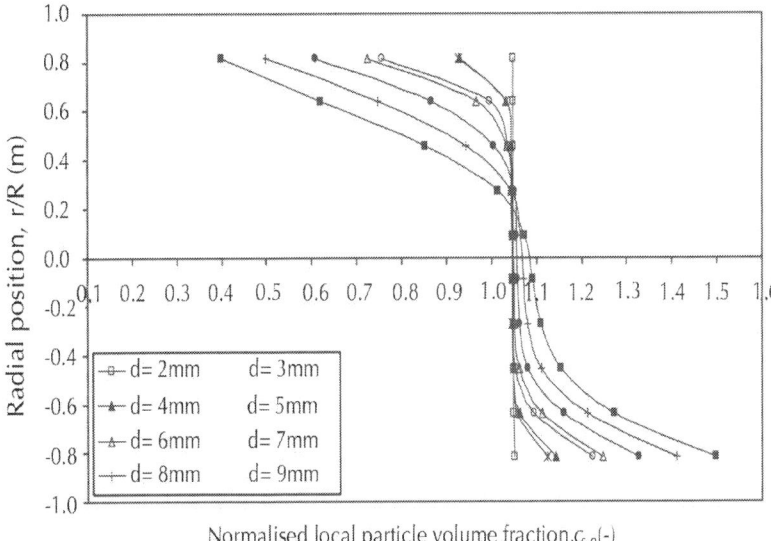

Figure 8: Effect of particle diameter on normalised particle concentration profile: k=0.16 Pa sn; n=0.81; ρ_s=1020 kg m^{-3}; C$_s$=0.25; u$_m$=125 mm s^{-1}.

Effect on Pressure Drop

In addition to friction arising from the flow of the carrier fluid, the presence of the particles introduces three additional sources of friction: particle–particle, particle–fluid, and particle–wall friction, all of which are influenced by the surface area and, therefore, the diameter of the particles. The CFD model predicts a decrease in pressure drop as particle diameter increases, as shown in Fig. 9 representing the percentage increase in pressure drop relative to that incurred by the carrier fluid flowing alone at the same flowrate. Asd increases at a constant particle concentration, the total surface area of the solid phase decreases, thus reducing particle–particle, particle–fluid, and particle–wall friction, and hence the observed decrease in pressure drop. Such a reduction in pressure drop, however, diminishes in significance at larger values of d, probably due to the increased propensity of particle settling which tends to increase particle-wall friction.

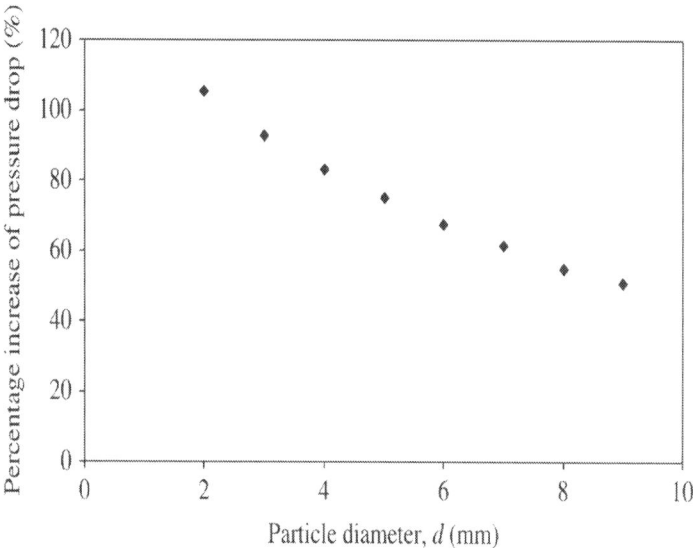

Figure 9: Effect of particle diameter on percentage increase in pressure drop relative to fluid flowing alone: k=0.16 Pa sn; n=0.81;ρ_s=1020 kg m^{-3}; C_s=0.25; u_m=125 mm s^{-1}.

Particle Concentration

The solids concentration was varied from 5% to 40% v/v while holding other parameters constant, as shown in Table 1. Significant effects were observed, as discussed below.

Effect on Solid Phase Velocity Profile

The solid phase velocity profiles predicted by CFD at different C_s values are presented in Fig. 10. The results show that increasing the particle concentration gives rise to increased blunting of the solid phase velocity profile, thus lowering the maximum velocity of particles near the pipe centre and raising the velocity of particles flowing near the wall, in agreement with experimental findings (Fairhurst et al., 2001; Fairhurst, 1998).

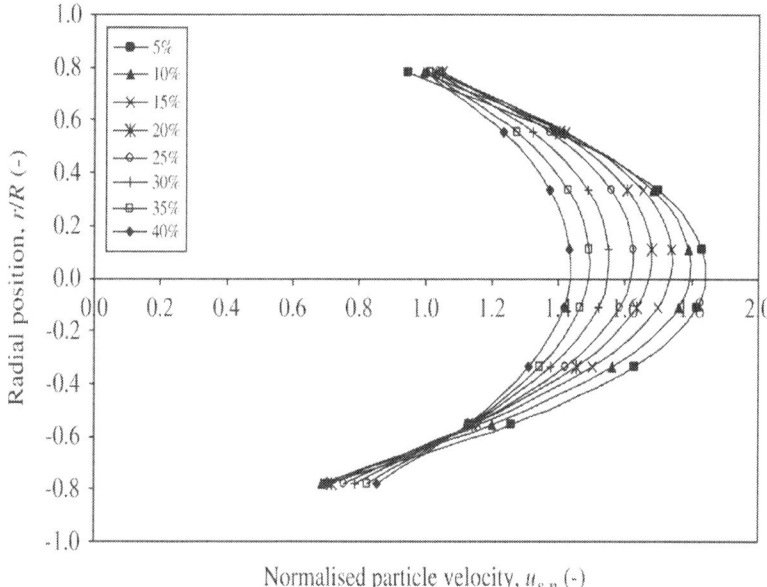

Figure 10: Effect of particle concentration on normalised solid-phase velocity profile: k=0.16 Pa sn; n=0.81; ρ_s=1020 kg m^{-3}; d=4 mm;u$_m$=125 mm s^{-1}.

Effect on Carrier Fluid Velocity Profile

Results in Fig. 11 indicate that increasing the solids concentration has a similar flattening effect on the velocity profile of the carrier fluid as it has on the solid phase velocity profile. As pointed out above, the fluid layer near the bottom wall of the pipe has a non-zero velocity that increases with solids concentration. This is due to the increasing velocity of the solid particles at higher C$_s$ values (Fig. 10), which disrupt the fluid boundary layer at the wall and generate a significant fluid slip. It is interesting to note that at C$_s$=0.4, for example, the fluid velocity at the bottom wall reaches a value about 20% of the maximum fluid velocity.

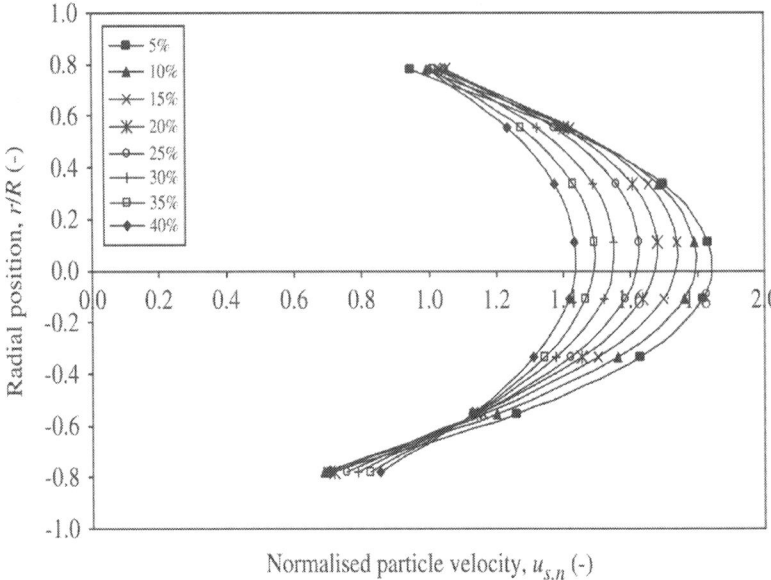

Figure 11: Effect of particle concentration on normalised carrier-fluid velocity profile: k=0.16 Pa sn; n=0.81; ρ_s=1020 kg m^{-3}; d=4 mm;u$_m$=125 mm s^{-1}.

Effect on Particle Concentration Profile

The normalised concentration profile of the solid particles is plotted as a function of solids concentration inFig. 12. The radial particle distribution becomes increasingly more uniform at higher solids loadings. At 40% v/v the concentration profile is almost flat across the entire pipe cross-section. As the maximum packing concentration is approached, no space is available for particles to settle out, thus leading to a uniform concentration profile. Experimental particle tracks determined by PEPT have led to the same observations and the identification of a "capsule flow" regime at such high particle concentrations (Barigou et al., 2003; Fairhurst et al., 2001). At lower C_s values, the central core region becomes more and more asymmetric around the centreline as the local particle concentration in the top half of the pipe declines.

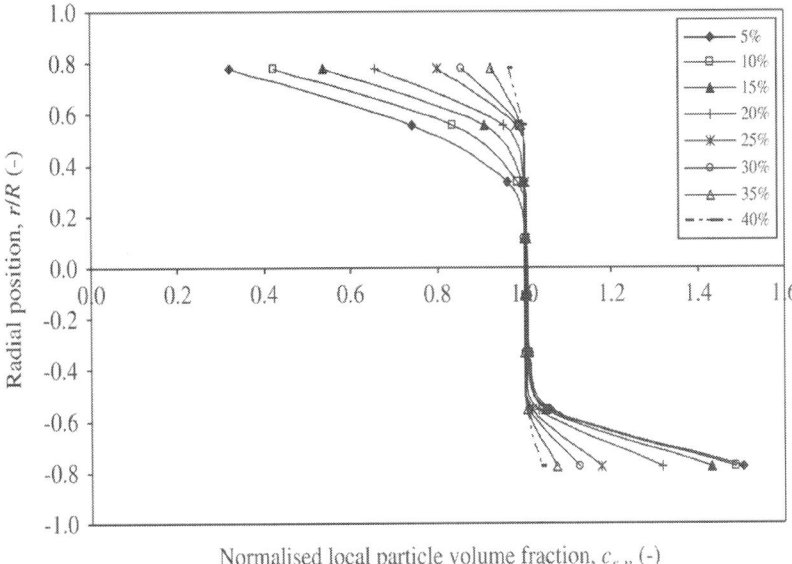

Normalised local particle volume fraction, $c_{s,n}$ (-)

Figure 12: Effect of particle concentration on normalised particle concentration profile: k=0.16 Pa sn; n=0.81; ρ_s=1020 kg m^{-3}; d=4 mm;u$_m$=125 mm s^{-1}.

Effect on Pressure Drop

As discussed above, the presence of solid particles in the carrying medium increases the pressure drop incurred due to particle–particle, particle–fluid and particle–wall interactions. Friction losses due to these interactions increase as the solids concentration increases, the number of particles increases and, hence, the total surface area of the solid phase increases. The percentage increase in pressure drop relative to that incurred by the carrier fluid flowing alone at the same flowrate was obtained from the CFD simulations at different particle concentrations, and the results are plotted in Fig. 13. There is a steep rise in solid–liquid pressure drop as a function of C_s; an increase in solids loading from 5% to 40% v/v results in about a tenfold increase in pressure drop. This represents a substantial rise in pressure drop compared to the carrier fluid flowing alone.

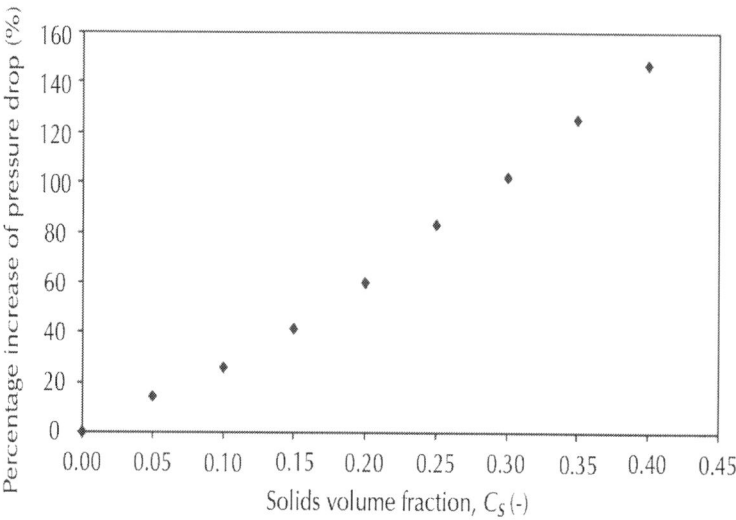

Figure 13: Effect of particle concentration on percentage increase in pressure drop relative to fluid flowing alone: k=0.16 Pa sn; n=0.81; $_s$=1020 kg m^{-3}; d=4 mm; u$_m$=125 mm s^{-1}.

Mixture Velocity

Varying the mean mixture velocity between 25 and 125 mm s^{-1} whilst the other flow conditions were kept constant as shown in Table 1, had little effect on the normalised solid phase and liquid phase velocity profiles, normalised particle concentration profile, and increase of pressure drop relative to the carrier fluid flowing alone. The solid phase velocity profile results are consistent with the experimental PEPT measurements obtained by Fairhurst (1998).

Rheological Properties

The effects of varying the rheological parameters of the shear-thinning carrier fluid were investigated, as follows.

Flow Behaviour Index, N

The flow behaviour index, n, indicates the extent of departure from Newtonian behaviour and is a measure of the degree of shear thinning of the fluid. This parameter was varied in the range 0.6–0.9 whilst all the other parameters were held constant at the values shown in Table 1. Note that the case of n>1 has already been addressed above (Section 5.1.1) in the context of the influence of shear thickening on the fluid and solid-phase velocity profiles.

Effect on Carrier Fluid Velocity Profile

The effects of the flow behaviour index on the velocity profile of the carrier fluid in the suspension are depicted in Fig. 14. There is a gradual flattening of the fluid velocity distribution as n decreases, i.e. as shear thinning increases. There is a certain degree of asymmetry caused by the slip layer at the bottom of the pipe, as discussed above. The velocity of this slip layer increases somewhat as n decreases. The velocity profile of the carrier fluid is compared to that of the fluid flowing alone in Fig. 15; the presence of the particles has a significant flattening effect on the fluid velocity profile approximately equal for all n values, where the maximum fluid velocity is reduced by some 20%.

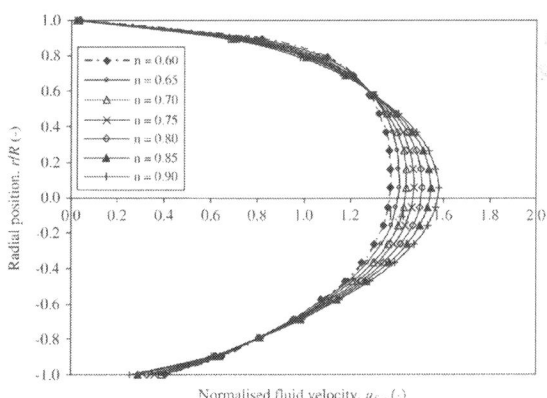

Figure 14: Effect of flow behaviour index on normalised carrier-fluid velocity profile: k=0.15 Pa sn; ρ_s=1020 kg m^{-3}; C$_s$=0.30; d=4 mm; u$_m$=125 mm s^{-1}.

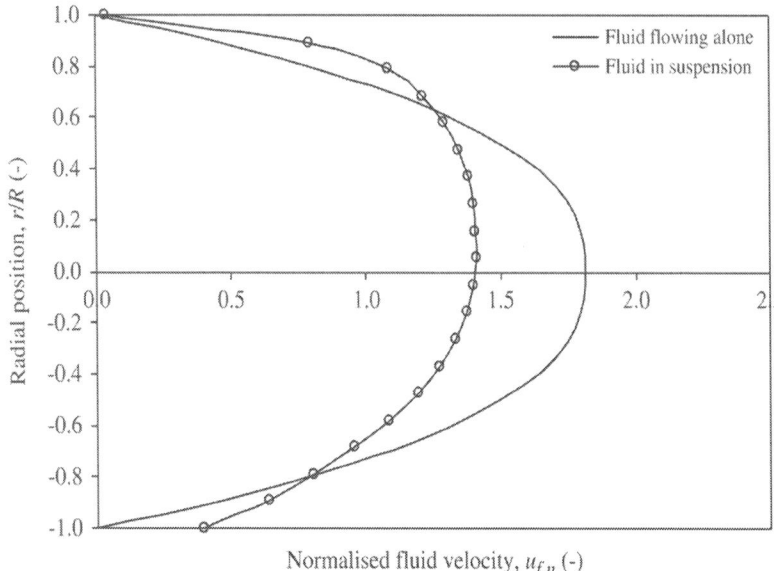

Figure 15: Effect of solid phase on normalised carrier-fluid velocity profile: k=0.15 Pa sn; n=0.65; ρ_s=1020 kg m^{-3}; C_s=0.30; d=4 mm; u_m=125 mm s^{-1}.

Effect on Solid-Phase Velocity Profile

The CFD-predicted velocity profiles presented in Fig. 16 show that particles in the top half of the pipe travel faster than those in the bottom half of the pipe, with the maximum occurring slightly above the centreline. This asymmetry in the velocity profile is greater for more shear thinning fluids, i.e. for lower values of n. Enhanced shear thinning also leads to a blunter solid-phase velocity profile as the fluid velocity profile becomes flatter (Fig. 14). However, similar to the velocity profile of the carrier fluid, the effects are not large for the practical range of n values investigated here: reducing n from 0.9 to 0.6 leads to only ~12% reduction in the maximum particle velocity.

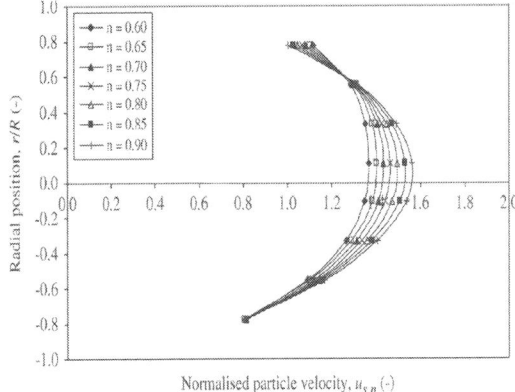

Figure 16: Effect of flow behaviour index on normalised solid-phase velocity profile: k=0.15 Pa sn; ρ_s=1020 kg m^{-3}; C$_s$=0.30; d=4 mm;u$_m$=125 mm s^{-1}.

Effect on Concentration Profile

Simulations showed that n had little effect on the particle concentration profile within the range of values studied, i.e. n=0.6–0.9.

Effect on Pressure Drop

The increase in pressure drop in solid–liquid flow relative to the single-phase pressure drop was found to be independent of n. It may therefore be inferred that the increase in pressure drop due to the presence of the solid phase is independent of the non-Newtonian behaviour of the carrier fluid so long as an effective viscosity can be defined for the fluid. This explains the good agreement obtained between the CFD predictions of pressure drop for mixtures with a non-Newtonian carrier fluid and the predictions of the semi-empirical correlation of Rasteiro et al. (1993), originally derived for a Newtonian fluid (see Table 2).

Consistency Index, K

The consistency index, k, was varied in the range 0.15–20 Pa sn with all other variables kept constant (Table 1). No significant effects were observed on the normalised solid phase and liquid phase velocity profiles, normalised particle concentration profile, and percentage increase of pressure drop relative to single-phase pressure drop.

Vertical Down-Flow

The effect of flow orientation was studied by considering vertical down-flow in the same 45 mm diameter pipe. The range of parameters investigated is shown in Table 1. CFD results showed that varying the particle diameter in the range 2–8 mm had little effect on the normalised solid-phase and liquid-phase velocity profiles and normalised concentration profile in vertical flow. The distribution of particles was remarkably uniform over the pipe cross-section. Increasing the particle diameter from 2 to 8 mm resulted in only ~10% reduction in pressure drop compared to single-phase fluid flow.

Solid concentration (varied within 5–30% v/v) also had no effect on the normalised solid-phase and liquid-phase velocity profiles and normalised concentration profile. The solid–liquid pressure drop, however, was substantially affected, as shown in Fig. 17. Increasing solids concentration led to an increased pressure loss due to increased particle–particle, particle–fluid and particle–wall interactions.

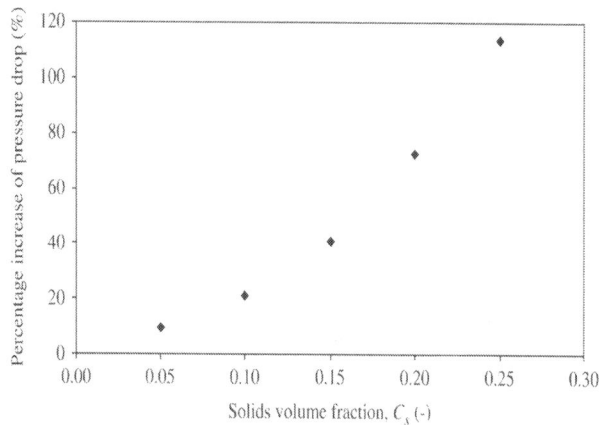

Figure 17: Effect of particle concentration in vertical down-flow on percentage increase in pressure drop relative to fluid flowing alone: k=0.16 Pa sn; n=0.81; ρ_s=1020 kg m^{-3}; d=4 mm; u$_m$=66 mm s^{-1}.

CONCLUSIONS

A parametric study of the horizontal and vertical flow of coarse nearly-neutrally solid particles in power law carrier fluids was conducted using an Eulerian–Eulerian two-fluid CFD model. The CFD-predicted solid-phase velocity profiles were successfully validated using experimental PEPT measurements, whilst the computed solid–liquid pressure drop was validated using correlations from the literature. The model was used to investigate the effects of particle diameter and concentration, mean flow velocity, and rheological properties of the carrier fluid on the solid-phase and liquid-phase velocity profiles, particle concentration profile, and pressure drop.

Except for the smaller particles, the velocity profile of the solid phase exhibited a significant degree of asymmetry which increased with particle size due to the increased propensity of particle settling. Also, with increasing particle size, the maximum solids velocity decreased and the position at which this velocity occurred shifted upwards above the centreline. Particles generally travelled faster than the mean mixture velocity, and while the maximum particle velocity was considerably less than twice the mean mixture velocity in shear-thinning carrier fluids, it exceeded this value in strongly shear thickening fluids ($\sim$$n$>1.75). Larger particles also caused significant blunting and asymmetry in the liquid-phase velocity profile. Whereas the particle concentration profile was nearly uniform for smaller particles, indicating a pseudo-homogeneous flow, increasing the particle diameter distorted the concentration profile due to enhanced settling. The solid–liquid pressure drop declined as particle diameter increased.

At higher solid concentrations, the solid-phase and liquid-phase velocity profiles became flatter and the particles were radially more uniformly distributed. Increasing the concentration also increased the pressure drop incurred considerably as the friction losses due to particle interactions increased. Increasing the mean flow velocity had no influence on the normalised velocity profile of either phase, the normalised particle concentration profile, or the pressure drop increase relative to the carrier fluid flowing alone.

The fluid consistency index did not have any influence on the normalised velocity profiles, normalised particle distribution, or pressure drop increase relative to single-phase flow. However, more shear thinning, i.e. lower n, induced a gradual flattening of the velocity profile of the solid phase and of the liquid-phase relative to the fluid flowing alone. No significant effects were observed on the normalised particle concentration profile or the increase in pressure drop compared to single-phase flow.

In vertical down-flow, particle radial distribution was remarkably uniform under all conditions investigated. The normalised solid-phase and liquid-phase velocity profiles and normalised particle concentration profile were independent of particle size and concentration under the conditions studied. However, increasing particle concentration resulted in a significant rise in pressure drop relative to single-phase flow.

REFERENCES

1.   Barigou, M., 2004. Particle tracking in opaque mixing systems: an overview of the capabilities of positron emission tomography. Chemical Engineering Research & Design 82 (A9), 1258–1267.

2.   Barigou, M., Fairhurst, P.G., Fryer, P.J., Pain, J.-P., 2003. Concentric flow regime of solid–liquid food suspensions: theory and experiment. Chemical Engineering Science 58, 1671–1686.

3.   Bouillard, J.X., Lyczkowski, R.W., Gidaspow, D., 1989. Porosity distribution in a fluidised bed with an immersed obstacle. A.I.Ch.E. Journal 35, 908–922.

4.   Brown, N.P., Heywood, N.I., 1991. Slurry Handling: Design of Solid–Liquid Systems. Elsevier Science Publishing Ltd.,

5.   Chakrabandhu, K., Singh, R.K., 2005. Rheological properties of coarse food suspensions in tube flow at high temperatures. Journal of Food Engineering 66, 117–128.

6.   Charles, M.E., Charles, R.A., 1971. In: Zandi, I. (Ed.), Advances in Solid–Liquid Flow and its Applications, Pergamon.

7.   Chhabra, R.P., Richardson, J.F., 1985. Hydraulic transport of coarse particles in viscous Newtonian and non-Newtonian media in a horizontal pipe. Chemical Engineering Research & Design 63, 390–397.

8.   Chhabra, R.P., Richardson, J.F., 1999. Non-Newtonian Flow in the Process Industries: Fundamentals and Engineering Applications. Butterworth-Heinemann, Ding, J., Gidaspow, D., 1990. A bubbling fluidization model using kinetic theory of granular flow. A.I.Ch.E. Journal 36 (4), 523–538.

9.   Duckworth, R.A., Pullum, L., Lockyear, C.F., 1983. The hydraulic transport of coarse coal at high concentration. Journal of Pipelines 3, 251–265.

10.   Duckworth, R.A., Pullum, L., Addie, G.R., Lockyear, C.F., 1986. Pipeline transport of coarse materials in a non-Newtonian carrier fluid. Hydrotransport 10, Paper C2, pp. 69–88.

11.   Durand, R., 1952. "The hydraulic transportation of coal and other materials in pipes," Colloquium of National Coal Board, London. Cited in Wasp, E.J., Kenny, J.P., Ghandi, R.L., 1979. Solid–liquid flow: slurry pipeline transportation. Trans Tech Publications, Clausthal, Germany.

12.   Eesa, M., Barigou, M., 2008. Horizontal laminar flow of coarse nearly-neutrally buoyant particles in non-Newtonian conveying fluids: CFD and PEPT experiments compared. International Journal of Multiphase Flow 34 (11), 997–1007.

13.   Enwald, H., Peirano, E., Almstedt, A., 1996. Eulerian two-phase flow theory applied to fluidization. International Journal of Multiphase Flow 22, 21–66.

14.   Fairhurst, P.G., 1998. Contribution to the study of the flow behaviour of large nearly neutrally buoyant spheres in non-Newtonian media: application to HTST processing. Ph.D. Thesis, Université de Technologie de Compiègne, France.

15.   Fairhurst, P.G., Barigou, M., Fryer, P.J., Pain, J.-P., Parker, D.J., 2001. Using positron emission particle tracking (PEPT) to study nearly neutrally buoyant particles in high solid fraction pipe flow. International Journal of Multiphase Flow 27, 1881–1901.

16.   Fregert, J., 1995. Velocity and concentration profiles for a laminar flow for a fluid containing large spheres in a horizontal pipe. Ph.D. Thesis, Lund University, Sweden.

17.   Ghosh, T., Shook, C.A., 1990. In: Liu, H., Round, G.F. (Eds.), Freight Pipelines, Hemisphere.

18. Gidaspow, D., 1994. Multiphase Flow and Fluidisation: Continuum and Kinetic Theory Descriptions. Academic Press, New York.

19. Gradeck, M., Fagla, B.F.Z., Baravian, C., Lebouché, M., 2005. Technical note: experimental thermomechanic study of Newtonian and non-Newtonian suspension flows. International Journal of Heat and Mass Transfer 48, 3477–3769.

20. Guer, Y.L.E., Reghem, P., Petit, I., Stutz, B., 2003. Experimental study of a buoyant particle dispersion in pipe flow. Transactions IChemE–Part A: Chemical Engineering Research & Design 81, 1136–1143.

21. He, Y.B., Laskowski, J.S., Klein, B., 2001. Particle movement in non-Newtonian slurries: the effect of yield stress on dense medium separation. Chemical Engineering Science 56, 2991–2998.

22. Hu, H.H., Patankar, N.A., Zhu, M.Y., 2001. Direct numerical simulations of fluid–solid systems using the arbitrary Lagrangian–Eulerian technique. Journal of Computational Physics 169, 427–462.

23. Kleinstreuer, C., 2003. Two-phase Flow: Theory and Applications. Taylor & Francis, London.

24. Krampa-Morlu, F.N., Bergstrom, D.J., Bugg, J.D., Sanders, R.S., Schaan, J., 2004.

25. Numerical simulation of dense coarse particle slurry flows in a vertical pipe. In: Proceedings of the Fifth International Conference on Multiphase Flow, ICMF'04, Paper No. 460.

26. Lareo, C., Fryer, P.J., Barigou, M., 1997. The fluid mechanics of two-phase solid–liquid food flows: a review. Transactions IChemE—Part C: Food and Bioproducts Processing 75, 73–105.

27. Legrand, A., Berthou, M., Fillaudeau, L., 2007. Characterization of solid–liquid suspensions (real, large non-spherical particles in non-Newtonian carrier fluid) flowing in horizontal and vertical pipes. Journal of Food Engineering 78, 345–355.

28. McCarthy, K.L., Kauten, R.J., Walton, J.H., 1996. Dynamics of fluid/particulate mixtures in tube flow. Short communication. Magnetic Resonance Imaging 14, 995–997.

29. McCarthy, K.L., Kerr, W.L., Kauten, R.J., 1997. Velocity profiles of fluid particulate mixtures using MRI. Journal of Food Process Engineering 20, 165–177.

30. Newitt, D.M., Richardson, J.F., Abbott, M., Turtle, R.B., 1955. Hydraulic conveying of solids in horizontal pipes. Transactions of the Institution of Chemical Engineers 33, 93–110.

31. Newitt, D.M., Richardson, J.F., Shook, C.A., 1962. Hydraulic conveying of solids in horizontal pipes. Part II: distribution of particles and slip velocities. In: Proceedings of the Interaction between Fluids and Particles, IChemE, London, pp. 87–100.

32. Rasteiro, M.G., Figueiredo, M.M., Franco, H., 1993. Pressure drop for solid/liquid flow in pipes. Particulate Science and Technology 11 (3–4), 147–155.

33. Shaw, C.T., 1992. Using Computational Fluid Dynamics. Prentice-Hall International (UK) Ltd., Shook, C.A., Roco, M.C., 1991. Slurry Flow: Principles and Practice. ButterworthHeimemann, Sumner, R.J., McKibben, M.J., Shook, C.A., 1990. Concentration and velocity distributions in turbulent vertical slurry flows. Ecoulements Solide–Liquide 2 (2), 33–42.

34. Thomas, D.G., 1965. Transport characteristics of suspensions. Part VIII. A note on the viscosity of Newtonian suspensions of uniform spherical particles. Journal of Colloid Science 20, 267–277.

35. Van Wachem, B.G.M., Almstedt, A.E., 2003. Methods for multiphase computational fluid dynamics. Chemical Engineering Journal 96, 81–98.

36. Wen, C.Y., Yu, Y.H., 1966. Mechanics of Fluidization. Chemical Engineering Progress Symposium Series 62, 100–108.

37. Zandi, I., Govatos, G., 1967. Heterogeneous flow of solids in pipelines. Journal of Hydraulics Division, Proceedings of the American Society of Civil Engineers, May, 93 (HY3), pp. 145–159.

Chapter 9

Long-Distance Axial Trapping with Focused Annular Laser Beams

Ming Lei, Ze Li., Shaohui Yan, Baoli Yao, Dan Dan, Yujiao Qi, Jia Qian, Yanlong Yang, Peng Gao, and Tong Ye

State Key Laboratory of Transient Optics and Photonics, Xi'an Institute of Optics and Precision Mechanics, Chinese Academy of Sciences, Xi'an, China

ABSTRACT

Focusing an annular laser beam can improve the axial trapping efficiency due to the reduction of the scattering force, which enables the use of a lower numerical aperture (NA) objective lens with a long working distance to trap particles in deeper aqueous medium. In this paper, we present an axicon-to-axicon scheme for producing parallel annular beams with the advantages of higher efficiency compared with the obstructed beam approach. The validity of the scheme is verified by the observation of a stable trapping of silica microspheres with relatively low NA microscope objective lenses (NA = 0.6 and 0.45), and the axial trapping depth of 5 mm is demonstrated in experiment.

INTRODUCTION

Since its first demonstration in 1986 by Ashkin et al. [1], optical tweezers has been serving as a powerful tool for microscopic trapping and manipulating, providing a stimulus to many research fields, such as in physics [2], biology [3] and colloid [4].

In an optical tweezers, stable trapping requires that the gradient force overcomes the scattering force, the former of which depends on the gradient of the intensity of the focused fields and the latter increases with increasing energy flow. Generally, stably axial trapping is more difficult than lateral trapping because of the intensively axial scattering force exerted by the axial energy flow in the focal region. As a result, the reduction of the axial scattering force is a key factor in stably axial trapping. Different approaches to improving axial trapping efficiency have been demonstrated to date. For larger particles, the use of higher-order Laguerre-Gaussian (LG) beam modes can improve axial trapping efficiency [5], [6]. But it should be noted that, although the LG beam is a hollow beam in intensity shape, it has a spiral phase distribution, which makes it have very different focusing property rather than other hollow beams with homogeneous phase distribution. Raktim Dasgupta et al. [7] utilized LG_{01} mode to trap silica microspheres at 200 μm axial distance. Rodrigo et al. [8], [9] demonstrated 3D trapping with long working distance via counter propagating light fields. Recently, radially polarized beam is shown to produce a vanishing axial Poynting vector component on the optical axis in the focal region, leading to a higher axial trapping efficiency in comparison with the linearly polarized beam or circularly polarized beam [10]–[14]. Double-ring radially polarized beams can further make the enhancement of the axial trapping efficiency [15]. Bowman [16] and Thalhammer et al. [17]. demonstrated a "macro-tweezers" approach, and the optical mirror trap was created after reflection of two holographically shaped collinear beams on a mirror. Although all these approaches are promising, their implementation demand some special optical elements such as spiral phase plate or spatially varying retarders, which limits their applicability. Ashkin [18]predicated that the use of an obstructed beam could increase the axial trapping efficiency of a dielectric particle since the annular intensity distribution enhanced the contribution of rays with a large angle of convergence, that would decrease the axial scattering force. Gu and Morrish[19] proved that Mie metallic particles were axially trapped with a centrally obstructed Gaussian (TEM_{00}-mode) beam focused by a high NA objective lens. Commonly, an opaque disk is used in the obstructed beam approach. This allows only those rays converging at large angles to be focused. The maximal axial trapping efficiency increases with the size of the center obstruction, but most of the incident light will be lost in this geometry. In addition, most of the axial trapping techniques developed so far utilize objective lenses with high numerical aperture (NA>1), which enable higher spatial resolution and axial trapping efficiency. However, high NA objective lenses are generally designed with a very short working distance (typically less than 0.2 mm), that means in many cases that the available space is too small to move the sample axially. High NA objectives also suffer from spherical aberrations when used for imaging in aqueous solutions. The spherical aberration introduced by the refractive difference between glass and water will produce a degradation of the imaging performance and inevitably limit the trapping depth in the aqueous medium [20], [21].

In this paper, we investigate theoretically the optical trapping efficiency on dielectric particles and make comparison of the axial and lateral trapping efficiencies for different widths of the annular beams by using vectorial diffraction theory. Long working distance objectives from Nikon Inc. and Edmund Optics Inc. (working distance are 8 mm and 13 mm, respectively) are used to focus the parallel annular laser beam to near diffraction-limited focal spot and realize the three-dimensional optical trapping. The parallel annular beam is generated by a telescopic pair of axicon lenses with transmittance of nearly 100%. The axial trapping depth of 5 mm is realized in the experiment. Theoretical calculations and experimental results are in good agreement.

THEORETICAL CALCULATION

In this section, we present the numerical results to show how an annular beam improve the axial trapping efficiency. The optical trapping force F on the particle can be calculated by vectorial diffraction theory based on the electromagnetic scattering theory [11], [22]. Instead of using the force F, we generally use a dimensionless quantity, trapping efficiency Q, to characterize how effective the optical trapping is. The trapping efficiency Q is defined, in component form, as

$$Q_i = \frac{\langle F_i \rangle}{(n_1 P/c)}, i = x, y, z,$$

(11)

where P is the power of the incident laser beam, c is the speed of light in vacuum, n_1 is the refractive index of surrounding medium, and $\langle F_i \rangle$ is the time-averaged trapping force. Throughout the paper, the wavelength is set to be λ = 491 nm, the refractive index of the particle is n_2 = 1.5 and that of the surrounding medium is n_1 = 1.33, the radius of the spherical particle is a = 2 μm. For the convenience of discussion, a parameter d, which can be regarded as the normalized width of the annular beams is introduced and defined as d = $(r_2 - r_1)/r_2$, where r_1 and r_2 are the inner radius and outer radius of the annular beam, respectively. It can be seen that with the increasing of the value of d, the width of the annulus increases accordingly.

The axial trapping efficiency Q_z is calculated based on the vectorial diffraction theory [23], [24].Figure 1 (a) shows the calculated curves of the axial trapping efficiencies of the annular beams with different d focused by a NA = 0.6 objective. For small values of d, the maximal backward trapping efficiency Q_{max} increases with increasing of d. This corresponds to the cases of d = 0.4 and 0.5. With further increasing of d, Q_{max}, however, decreases, as can be seen for d = 0.6 and 1.0 curves. This implies that there is a critical value of d, say d_0, for which Q_{0max} assumes its maximum. The occurrence of d_0 is a result of the competition between the scattering force and the gradient force in the focal region. A annular with small values of d can offer a higher portion of rays with large converging angle of focusing, leading to a reduction of the scattering force. However, a smaller value of d also results in the extent of the axial focal depth, which will reduce the axial gradient force. Therefore, it is important to find an optimal value of d in practice. From Figure 1(a) we can see that the maximal backward axial trapping efficiency at the optimal annulus (d = 0.5) has an improvement up to 49.2% in comparison with the TEM_{00} beam (d = 1).

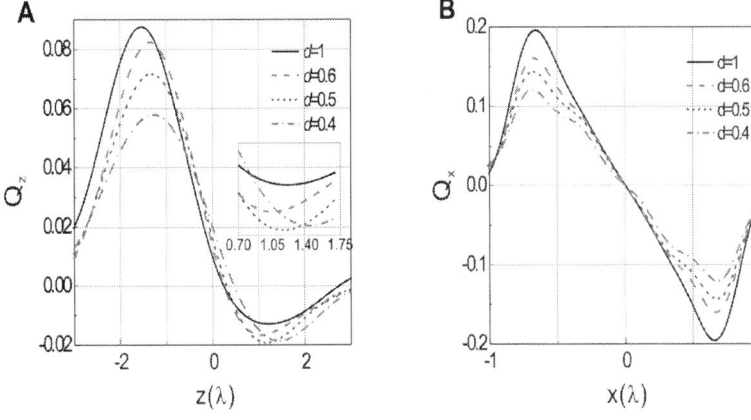

Figure 1: Calculated trapping efficiencies of a spherical particle by annular beams with different widths. A. Axial trapping efficiency, B. lateral trapping efficiency. The trapping wavelength is 491 nm, the sphere radius is 2 μm, and the numerical aperture of the objective is 0.6.

Figure 1(b) gives the lateral trapping efficiency for different values of d. All the while, the maximal lateral trapping efficiency falls with decreasing of the values of d. This is because that the gradient force is dominant due to the multiple refraction on the surface of the trapped dielectric particle [25], i.e., there will be much loss of lateral trapping force when d is smaller. However, the reduction of lateral trapping efficiency has little influence on stably lateral trapping in practice.

METHODS

Axicon (conical lens) is a well known wavefront division optical element to transform a Gaussian beam into a Bessel beam, and has already found applications in optical trapping [26]and multifunctional darkfield microscopy [27]. Shao et al. [28] utilized a 3-axicon approach to generate a three-dimensional ring-shaped trap. We adopt here a telescopic pair of axicons to transform a Gaussian beam into a parallel annular beam. As shown in Figure 2, a collimated Gaussian beam incident on the base of the first axicon is deviated cylindrically toward the optical axis due to refraction. All deviated rays independently propagate along different directions and form a hollow cone beam, and then incident on the conical surface of the second axicon with a ring-shape of intensity distribution. Two axicons with the same open angle are arranged tip to tip to ensure the output beam behind the second axicon parallel to the optical axis. Adjusting the relative distance between the two axicons corresponds to changing the diameter of the output parallel annular beam [29]. By adjusting the diameter of the output parallel annular beam to overfill the back aperture of the objective len, a near diffraction-limited focus can be obtained. Considering the light transmittance, the open angle γ of the axicon is usually designed very small (less than 10 degree), giving a good approximate calculation of the divergence angle $u \approx (n-1)\gamma$, where n is the refractive index of the axicon. We used two axicons from Thorlabs Inc. with the open angle $\gamma = 5°$ and the refractive index n = 1.5. One of the advantages of using the axicon-pair geometry is the very high transmittance, which can be nearly 100% in theory. Considering the transmittance of the antireflection coating of the two axicons, the total conversion efficiency can still reach to 90% in practice.

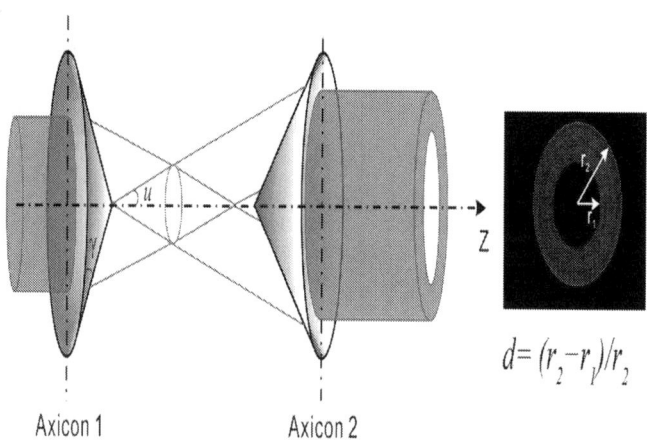

$$d = \left(r_2 - r_1\right)/r_2$$

Axicon 1   Axicon 2

Figure 2: Axicon-pair for generation of parallel annular beam.

Figure 3 shows the experimental layout of our trapping system. A diode-pumped solid-state laser (Calypso 491, Cobolt AB Inc., Sweden) working at wavelength of 491 nm is expanded by a telescope formed by Lens 1 and Lens 2. After beam expanding, the collimated beam passes through the telescopic pair of axicon to be transformed into a parallel annular beam. The parallel annular beam then passes through another telescope comprised of Lens 3 and Lens 4, and then is reflected by a dichroic mirror (short-pass 475 nm) into the back aperture of the long working distance 20X objective (EO M Plan HR, NA0.6, 13 mm working distance, Edmund Optics Inc., USA) or a 20X objective (Plan Fluor, NA0.45, 8 mm working distance, Nikon Inc., Japan). Sliding the Lens 4 will slightly change the divergence of the annular shaped beam and ensure that it can be focused on the sample exactly. The sample is mounted on a motorized XYZ stage (MP-285, Sutter Instrument Inc., Canada) that can be driven either manually or by a programmable software interface. A USB CCD camera (DMK 41BU02, 1280×960 pixels, 4.65 µm×4.65 µm pixel size, The Imaging Source Europe GmbH, Germany) is employed to record the trapping process, with an lens (50 mm, F/1.8, Nikon Inc, Japan) serving as a tube lens. The CCD camera allows imaging speed as high as 30 fps at a resolution of 1280×960 pixels. A long-pass filter (long-pass 505,) is inserted in front of the CCD camera to block the trapping laser beam. The maximal laser power delivered on the sample is about 10 mW. Silica microspheres (4 µm in diameter, Polysciences Inc., USA) immersed in water in a sample chamber are used as test samples.

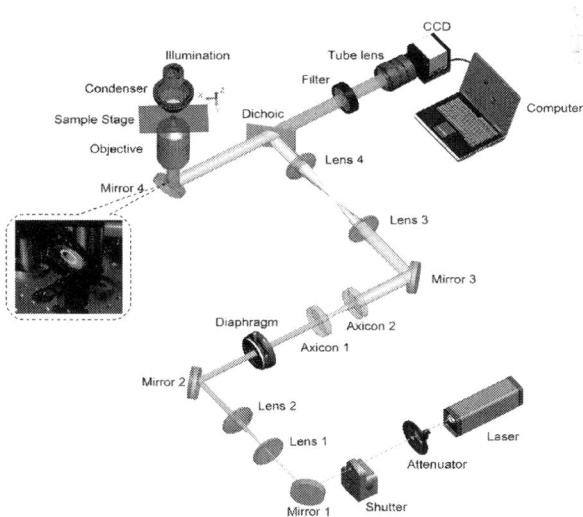

Figure 3: Experimental layout of the trapping system. The inset gives the annular beam shape captured on Mirror 4.

RESULTS AND DISCUSSION

The axial trapping force F_z on a trapped silica microsphere suspended in water was measured by observing the maximal translation speed of the motorized stage at which the particle fell out of the trap. The force is calculated by the Stokes law $F = 6\varpi\eta av$, where a is the radius of a trapped particle, v is the maximal translation speed, and η is the viscosity of the surrounding medium. The maximal axial trapping efficiency $Q_{z,max}$ is then calculated by Eq.(1). In the trapping experiment, the upward motion of the motorized stage is chosen, because the trap is weakest in that direction in the inverted microscope due to the opposing action between the scattering force and the gradient force. The experiment is repeated several times to reduce the error. The measured maximal axial trapping efficiency is 0.0104. Figure 4 (Video S1) demonstrates actually an axial movement of a trapped silica bead in 5 mm depth with a 20X/NA0.6 objective. The axial movement of the particle is controlled by a motorized stage with resolution of 200 nm, and the moving speed and position are controllable with the software.Figure 4A–F show the trapped bead at different axial planes. According to the working distance of the microscope objective, the maximal axial trapping depth can theoretically reach up to 13 mm.

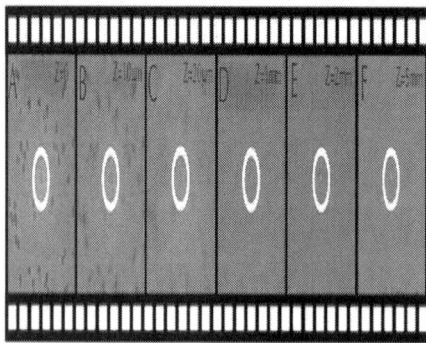

Figure 4: Trapping and moving a silica microsphere in 5 mm depth axially with the focused annular beam.A–F show the trapped microsphere at different axial planes. Using a long working distance microscope objective (20X/NA0.6) (Video S1).

As expected, the trapping force caused by the TEM_{00} Gaussian beam focused by the same objective failed to trap the same particle, and always pushed the particle out of the focus. This process is demonstrated in Figure 5 (Video S2).

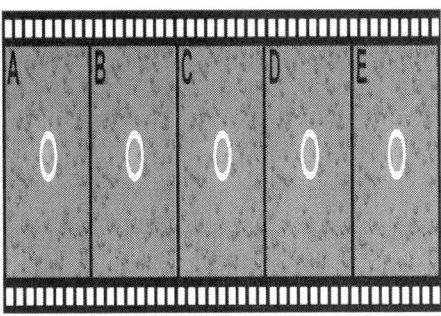

Figure 5: Pushing a silica microsphere out of the focus with the focused TEM-$_{00}$beam.Using a long working distance microscope objective (20X/NA0.6). (Video S2).

We also demonstrate that it is even possible to trap the particles by an objective with much lower NA(0.45), as shown in Figure 6 (Video S3).

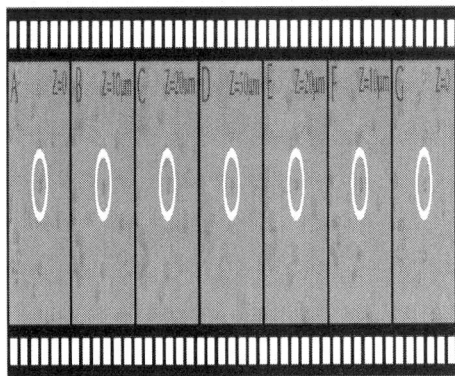

Figure 6: Trapping and moving a silica microsphere axially by an objective lens with much lower NA(0.45).A–D moving upward, E–G moving backward. (Video S3).

Gu et al. [19] proposed a method for three-dimensional optical trapping of metallic Mie particles using an obstructed laser beam based on the geometrical optics model. The maximal axial trapping efficiency increases with increasing of the size of the center obstruction in their calculation. However, the geometrical optics model ignores the light intensity distribution in the focal region, and can only be employed as a good approximation to compute the force on particles which are much larger than the wavelength of the incident light. We tried decreasing the width of the annular beam to observe the impact on the axial trapping efficiency, but contrary to the calculation based on geometrical optics model, the axial trapping efficiency will decrease as the annular width keeps decreasing after the optimal value d = 0.5. We think this is because that the width of the annular beam has an impact on the intensity distribution near the focal region. Figure 7 gives the simulated lateral and axial intensity distributions for the annular beam with different widths focused by a NA = 0.6 objective. It can be seen that the lateral intensity distribution keeps almost unaltered with different widths. While for axial direction, the thin annular light field will lead to the focus to be stretched a lot along the axial direction, thereby reducing the axial trapping efficiency. This phenomenon was also mentioned by Kitamura et al. [30] The theoretical calculation of optical trapping force in our simulation is based on the vectorial diffraction theory, which provides a more rigorous solution of the scattering field and is not limited by the size and shape of particles in contrast with the geometrical optics model.

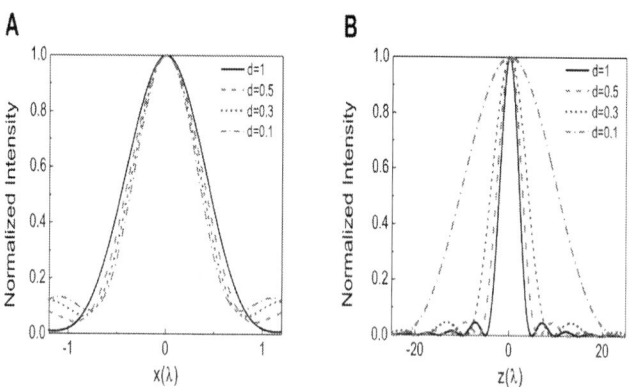

Figure 7: Simulated intensity distributions of the focused annular beam with different widths.A. Lateral intensity distribution, B. axial intensity distribution. The numerical aperture of the objective is 0.6.

Trapping stiffness is a common parameter to quantify the stability of an optical tweezers system. It can be measured and calculated by using the probability distribution method. Generally, there are two main routes to measure the motion of trapped particles, i.e., quadrant photodiode (QPD) [31] and video-based particle tracking [32]. Due to the fluctuation of trapped particles is normally in the order of nanometers, such measurement typically needs oil-immersion objective with high magnification. In our presented system, a 20X objective is used to trap 4 μm-diameter particles. The resolution of our imaging system is 230 nm/pixel approximately, which is barely enough for the position measurement algorithm. From our trapping videos, we just estimate that the lateral stiffness of our tweezers system is about 3×10^{-6} N/m @ 20 mW. For integrity, we also applied a trapping force calibration technique as described above. The measured maximal axial trapping efficiency is consistent with the theoretical calculation.

Optical trapping of using large NA objectives has some limitations such as extremely short working distance, narrow field of view, and tight focusing with high power density, which might cause heating and optical damage to the biological specimen. In some applications, a large field of view and long working distance are highly desirable. Long working distance objectives are particularly useful for industrial inspection such as wafer probing and flaw detection. The utilization of such kind of objective in optical trapping may open the possibility of exploring large volume with optical tweezers. Due to the long distance imaging capability, the axicon-pair-based optical tweezers could also be a useful tool in various biological researches, for example, examining specimens in vitro through thick glass walls, where the objective lens must be protected against environmental hazards such as heat, vapors, and volatile chemicals by a thick coverslip.

CONCLUSIONS

We have proposed a long distance axial trapping approach with focused annular laser beam based on a telescopic pair of axicon. The optical trapping efficiencies on dielectric particles for different widths of the annular beams have been calculated by using vectorial diffraction theory. By trapping silica microspheres suspended in water, we have demonstrated that the system has the advantages of higher efficiency and longer trapping range over the conventional axial trapping geometry.

AUTHOR CONTRIBUTIONS

Conceived and designed the experiments: ML BY. Performed the experiments: ML ZL. Analyzed the data: ZL SY DD. Contributed reagents/materials/analysis tools: YQ JQ YY PG TY. Wrote the paper: ML ZL SY.

REFERENCES

1. Ashkin A, Dziedzic JM, Bjorkholm JE, Chu S (1986) Observation of a single-beam gradient force optical trap for dielectric particles. Opt. Lett 11: 288–290. doi: 10.1364/ol.11.000288

2. Higuchi T, Pham QD, Hasegawa S, Hayasaki Y (2011) Three-dimensional positioning of optically trapped nanoparticles. Appl. Opt 50: H183–H188. doi: 10.1364/ao.50.00h183

3. Carmon G, Feingold M (2011) Rotation of single bacterial cells relative to the optical axis using optical tweezers. Opt. lett 36: 40–42. doi: 10.1364/ol.36.000040

4. Gutsche C, Elmahdy MM, Kegler K, Semenov I, Stangner T, et al. (2011) Micro-rheology on (polymer-grafted) colloids using optical tweezers. J. Phys.: Condens. Matter 23: 184114. doi: 10.1088/0953-8984/23/18/184114

5. Simpson NB, McGloin D, Dholakia K, Allen L, Padgett MJ (1998) Optical tweezers with increased axial trapping efficiency. Journal of Modern Optics 45: 1943–1949. doi: 10.1080/09500349808231712

6. O'Neil AT, Padgett MJ (2001) Axial and latteral trapping efficiency of Laguerre-Gaussian modes in inverted optical tweezers. Opt. Commun 193: 45–50. doi: 10.1016/s0030-4018(01)01198-1

7. Dasgupta R, Verma RS, Ahlawat S, Chaturvedi D, Gupta PK (2011) Long-distance axial trapping with Laguerre–Gaussian beams. Appl. Opt 50: 1469–1476. doi: 10.1364/ao.50.001469

8. Rodrigo PJ, Daria VR, Glückstad J (2004) Real-time three-dimensional optical micromanipulation of multiple particles and living cells. Opt. Lett 29: 2270–2272. doi: 10.1364/ol.29.002270

9. Rodrigo PJ, Daria VR, Glückstad J (2005) Four-dimensional optical manipulation of colloidal particles. Appl. Phys. Lett 86: 074103. doi: 10.1063/1.1866646

10. Zhan QW (2004) Trapping metallic Rayleigh particles with radial polarization. Opt. Express 15: 3377–3382. doi: 10.1364/opex.12.003377

11. Yan SH, Yao BL (2007) Radiation forces of highly focused radially polarized beam on spherical particles. Phys. Rev A 76: 053836. doi: 10.1103/physreva.76.053836

12. Michihata M, Hayashi T, Takaya Y (2009) Measurement of axial and transverse trapping stiffness of optical tweezers in air using a radially polarized beam. Appl. Opt 48: 6143–6151. doi: 10.1364/ao.48.006143

13. Iglesias I, Saenz JJ (2012) Light spin forces in optical traps: comment on "Trapping metallic Rayleigh particles with radial polarization". Opt. Express 20: 2832–2834. doi: 10.1364/oe.20.002832

14. Donato MG, Vasi S, Sayed R, Jones PH, Bonaccorso F, et al. (2012) Optical trapping of nanotubes with cylindrical vector beams. Opt. Lett 37: 3381–3383. doi: 10.1364/ol.37.003381

15. Yao BL, Yan SH, Ye T, Zhao W (2010) Optical trapping of double-ring radially polarized beam with improved axial trapping efficiency. Chin. Phys. Lett 27: 108701. doi: 10.1088/0256-307x/27/10/108701

16. Bowman R, Jesacher A, Thalhammer G, Gibson G, Ritsch-Marte M, et al. (2011) Position clamping in a holographic counterpropagating optical trap. Opt. Express 19: 9908–9914. doi: 10.1364/oe.19.009908

17. Thalhammer G, Steiger R, Bernet S, Ritsch-Marte M (2011) Optical macro-tweezers: trapping of highly motile micro-organisms. J Opt 13: 044024. doi: 10.1088/2040-8978/13/4/044024

18. Ashkin A (1992) Forces of a single-beam gradient laser trap on a dielectric sphere in the ray optics regime. Biophy J 61: 569–582. doi: 10.1016/s0006-3495(92)81860-x

19. Gu M, Morrish D (2002) Three-dimensional trapping of Mie metallic particles by the use of obstructed laser beams. J Appl. Phys 91: 1606–1612. doi: 10.1063/1.1428801

20. Reihani SNS, Charsooghi MA, Khalesifard HR, Golestanian R (2006) Efficient in-depth trapping with an oil-immersion objective lens. Opt. Lett 31: 766–768. doi: 10.1364/ol.31.000766

21. Vermeulen KC, Wuite GJL, Stienen GJM, Schmidt CF (2006) Optical trap stiffness in the presence and absence of spherical aberrations. Appl. Opt 45: 1812–1819. doi: 10.1364/ao.45.001812

22. Yan SH, Yao BL (2007) Transverse trapping forces of focused Gaussian beam on ellipsoidal particles. J Opt. Soc. Am B 24: 1596–1602. doi: 10.1364/josab.24.001596

23. Ganic D, Gan XS, Gu M (2004) Exact radiation trapping force calculation based on vectorial diffraction theory. Opt. Express 12: 2670–2675. doi: 10.1364/opex.12.002670

24. Ganic D, Gan XS, Gu M (2005) Optical trapping force with annular and doughnut laser beams based on vectorial diffraction. Opt. Express 13: 1260–1265. doi: 10.1364/opex.13.001260

25. Gu M, Morrish D, Ke PC (2000) Enhancement of transverse trapping efficiency for a metallic particle using an obstructed laser beam. Appl. Phys. Lett 77: 34–36. doi: 10.1063/1.126868

26. Garces-Chavez V, McGloin D, Melville H, Sibbett W, Dholakia K (2002) Simultaneous micromanipulation in multiple planes using a self-reconstructing light beam. Nature 419: 145–147. doi: 10.1038/nature01007

27. Lei M, Yao BL (2008) Multifunctional darkfield microscopy using an axicon. J Biomed. Opt 13: 044024. doi: 10.1117/1.2960019

28. Shao B, Esener SC, Nascimento JM, Berns MW, Botvinick EL, et al. (2006) Size tunable three-dimensional annular laser trap based on axicons. Opt. Lett 31: 3375–3377. doi: 10.1364/ol.31.003375

29. Golub I, Tremblay R (1990) Light focusing and guiding by an axicon pair generated tubular light beam. J Opt.Soc. Am B 7: 1264–1267. doi: 10.1364/josab.7.001264

30. Kitamura K, Sakai K, Noda S (2010) Sub-wavelength focal spot with long depth of focus generated by radially polarized, narrow-

width annular beam. Opt. Express 18: 4518–4525. doi: 10.1364/oe.18.004518

31. Sørensen KB, Flyvbjerg H (2004) Power spectrum analysis for optical tweezers. Rev. Sci. Instrum 75: 594–612. doi: 10.1063/1.1645654

32. Otto O, Czerwinski F, Gornall JL, Stober G, Oddershede LB, et al. (2010) Real-time particle tracking at 10,000 fps using optical fiber illumination. Opt. Express 18: 22722–22733. doi: 10.1364/oe.18.022722

Citations

CHAPTER 1

J. Sousa, C. Sodré, A. Lima and S. Neto, "Numerical Analysis of Heavy Oil-Water Flow and Leak Detection in Vertical Pipeline," Advances in Chemical Engineering and Science, Vol. 3 No. 1, 2013, pp. 9-15. doi: 10.4236/aces.2013.31002.

CHAPTER 2

Nakao, M. , Kato, T. , Oowaku, T. , Sakuma, H. and Kagawa, T. (2014) Generation of Arbitrary Pressure Pulsation of Wide Frequency Range for Flow Meter Testing in a Laminar Gas Pipeline. Journal of Flow Control, Measurement & Visualization, 2, 125-137. Doi: 10.4236/jfcmv.2014.24015.

CHAPTER 3

Mainier, F. and Guimarães, P. (2014) Use of Corrosion Inhibitor in Solid Form to Prevent Internal Corrosion of Pipelines and Acidification Process. Journal of Materials Science and Chemical Engineering, 2, 1-6. doi:10.4236/msce.2014.25001.

CHAPTER 4

M. Hameed and M. Jawad, "Hydrodynamics of Liquid Film in Helical Tubes," Advances in Chemical Engineering and Science, Vol. 2 No. 1, 2012, pp. 74-81. doi: 10.4236/aces.2012.21009.

CHAPTER 5

G. Silva, V. Filho and L. Bahiense, "Computational Tools to Support Ethanol Pipeline Network Design Decisions," American Journal of Operations Research, Vol. 3 No. 1, 2013, pp. 1-16. doi: 10.4236/ajor.2013.31001.

CHAPTER 6

G. Tseropoulos, Y. Dimakopoulos, J. Tsamopoulos, G. Lyberatos, On the flow characteristics of the conical Minoan pipes used in water supply systems, via computational fluid dynamics simulations, Journal of Archaeological Science, Volume 40, Issue 4, April 2013, Pages 2057-2068, ISSN 0305-4403, http://dx.doi.org/10.1016/j.jas.2012.11.025.

CHAPTER 7

Wenhao Pu, Changsui Zhao, Yuanquan Xiong, Cai Liang, Xiaoping Chen, Peng Lu, Chunlei Fan, Numerical simulation on dense phase pneumatic conveying of pulverized coal in horizontal pipe at high pressure, Chemical Engineering Science, Volume 65, Issue 8, 15 April 2010, Pages 2500-2512, ISSN 0009-2509, http://dx.doi.org/10.1016/j.ces.2009.12.025.

CHAPTER 8

M. Eesa, M. Barigou, CFD investigation of the pipe transport of coarse solids in laminar power law fluids, Chemical Engineering Science, Volume 64, Issue 2, January 2009, Pages 322-333, ISSN 0009-2509, http://dx.doi.org/10.1016/j.ces.2008.10.004.

CHAPTER 9

Lei M, Li Z, Yan S, Yao B, Dan D, et al. (2013) Long-Distance Axial Trapping with Focused Annular Laser Beams. PLoS ONE 8(3): e57984. doi:10.1371/journal.pone.0057984.

Index